Praise for *Everything and More*

"*Everything and More* is, in nearly every way, a gift. It's a thoughtful and witty 300-page testimonial to the qualities I never fully understood that mathematics possessed: Math is astonishing and full of 'shadowlands,' and—ultimately—stunning beauty."
—Anthony Doerr, *Boston Globe*

"Wallace is the perfect parachute buddy for a free fall into the mathematical and metaphysical abyss that is infinity."
—Dennis Lim, *Village Voice*

"Shockingly readable . . . a brilliant antidote both to boring math textbooks and to pop-culture math books that emphasize the discoverer over the discovery."
—*Booklist*, starred review

"Had he not pursued a career in literary fiction, it's not difficult to imagine Wallace as a historian of science, producing quirky and challenging volumes such as this every few years."
—*Publishers Weekly*

PUBLISHED TITLES IN THE GREAT DISCOVERIES SERIES

David Foster Wallace
Everything and More: A Compact History of ∞

Sherwin B. Nuland
The Doctors' Plague: Germs, Childbed Fever, and the Strange Story of Ignác Semmelweis

Michio Kaku
Einstein's Cosmos: How Albert Einstein's Vision Transformed Our Understanding of Space and Time

Barbara Goldsmith
Obsessive Genius: The Inner World of Marie Curie

Rebecca Goldstein
Incompleteness: The Proof and Paradox of Kurt Gödel

Madison Smartt Bell
Lavoisier in the Year One: The Birth of a New Science in an Age of Revolution

George Johnson
Miss Leavitt's Stars: The Untold Story of the Woman Who Discovered How to Measure the Universe

David Leavitt
The Man Who Knew Too Much: Alan Turing and the Invention of the Computer

William T. Vollmann

Uncentering the Earth: Copernicus and The Revolutions of the Heavenly Spheres

David Quammen

The Reluctant Mr. Darwin: An Intimate Portrait of Charles Darwin and the Making of His Theory of Evolution

Richard Reeves

A Force of Nature: The Frontier Genius of Ernest Rutherford

Dan Hofstadter

The Earth Moves: Galileo and the Roman Inquisition

Michael D. Lemonick

The Georgian Star: How William and Caroline Herschel Revolutionized Our Understanding of the Cosmos

FORTHCOMING TITLES

Lawrence M. Krauss

Quantum Man: Richard Feynman's Life in Science

BY DAVID FOSTER WALLACE

The Broom of the System

Girl with Curious Hair

Infinite Jest

A Supposedly Fun Thing I'll Never Do Again

Brief Interviews with Hideous Men

Everything and More: A Compact History of ∞

Oblivion

DAVID FOSTER WALLACE

Everything and More

A Compact History of ∞

ATLAS BOOKS

W. W. NORTON & COMPANY
NEW YORK · LONDON

Printed in the United States of America
First published as a Norton paperback 2004; reissued 2010

Manufacturing by LSC Harrisonburg
Production manager: Julia Druskin

Library of Congress Cataloging-in-Publication Data
Wallace, David Foster.
Everything and more : a compact history of infinity / David Foster Wallace.—1st ed.
p. cm. — (Great discoveries)
Includes bibliographical references.
ISBN 0-393-00338-8
1. Infinite—History. I. Title. II. Series.
QA9.W335 2003
511.3—dc21

2003011415

ISBN 978-0-393-33928-4 pbk.

W. W. Norton & Company, Inc., 500 Fifth Avenue,
New York, N.Y. 10110
www.wwnorton.com

W. W. Norton & Company Ltd.
15 Carlisle Street, London W1D 3BS

9 0

For my mother and father

οὐκ ὃ ἐν τῇ κεφαλῇ, ἀλλ’ ἐν ᾧ ἡ κεφαλή ἐστιν.

Everything and More

A Compact History of ∞

Introduction
Neal Stephenson

When I was a boy growing up in Ames, Iowa, I belonged to a Boy Scout troop whose adult supervision—consisting almost entirely of professors from the Iowa State University of Science and Technology—devised the following project for us to pursue when not occupied with dodgeball and clove hitches. One of the scouts' dads—an eminent professor of agricultural engineering—obtained, from a lab in his department, a sack of genetically identical corn kernels, carried them across campus, and handed them off to one of the other scouts' dads: a physicist employed by the Ames Laboratory. This was an offshoot of the Manhattan Project. The uranium enriched at Oak Ridge, and used in the first atomic bombs, had been refined from its ore by a process developed at Ames. Dad #2, who had been present at the startup of the world's first atomic pile in a racquetball court at the University of Chicago, carried the seeds into a hot room buried a couple of stories beneath one of the Ames Lab's buildings and handed it off to a mechanical arm that carried it behind a thick wall of yellowish lead-laced glass and set it down in the vicinity of something that was radioactive. After a certain amount of time had passed, he retrieved the irradiated seeds and brought them to the next meeting of our scout troop and distributed them to the boys. I distinctly remember looking at the kernels in the palm of my hand and noting that they had been washed with paint or ink of two or three different colors, and, though the color code was not explained to us (not, at least, before the expiration of my attention span), I caught the spoor of the scientific method and guessed that different batches had been exposed to greater or lesser amounts of radiation. In any case, we were directed to take these seeds home and plant them and water them. In a few weeks' time, we would bring the results to a meeting where two prizes would be handed out: one for the tallest, healthiest corn plant, the other for the weirdest mutation. And indeed we ended up with both: proud stalks that would do any Iowa farmer proud, and plants, in many cases quite beautiful, that were scarcely recognizable as belonging to the relevant taxonomic phylum. If anyone had asked us "do you imagine that other scout troops in other towns are doing anything remotely like this" we would, after some higher-brain activity, have guessed no. No one asked, however, and so our lower brains assimilated the whole scenario as normal, like playing catch and making s'mores.

I draw the reader's attention, in other words, to the phenomenon of the Midwestern American College Town, which, in a completely self-aware tip of the stylistic hat to David Foster Wallace, I will denominate the MACT. In 1960, when I was six months old, my parents and I moved to the archetypal, if somewhat larger-than-normal, MACT of Champaign-Urbana, Illinois, so that my father could get to work on his PhD. Two years later, when David Foster Wallace was six months old, his family moved to the same town on the same errand (his dad is a philosopher, mine an electrical engineer). He and I lived in the same MACT only until 1966, when my family moved to the smaller, but no less quintessential, MACT of Ames. I never met him, unless we happened to share a slide or a swingset in some Champaign-Urbana park. Each of us went to Massachusetts for higher education and then landed for a while in a different MACT: Iowa City in my case, Bloomington-Normal, Illinois, for DFW.

The irradiated-corn anecdote might have already said everything there is to say about the culture of the MACT, but, since DFW and I seem to have been MACT products all the way, there are a few particulars that might be worth drawing out in a more discursive manner. So here goes.

People who often fly between the East and West Coasts of the United States will be familiar with the region, stretching roughly from the Ohio to the Platte, that, except in anomalous non-flat areas, is spanned by a Cartesian grid of roads. They may not be aware that the spacing between roads is exactly one mile. Unless they have a serious interest in nineteenth-century Midwestern cartography, they can't possibly be expected to know that when those grids were laid out, a schoolhouse was platted at every other road intersection. In this way it was assured that no child in the Midwest would ever live more than $\sqrt{2}$ miles from a place where he or she could be educated. Secondary schools were presumably sited according to some less rigid scheme, and universities were generally doled out two to a state. According to a convention that obtains pretty consistently across all states west of Ohio, a given state, call it X, is allotted a "University of

X" and an "X State University." "University of X" has been a University, as opposed to a College, from its inception and generally houses all of the prestigious Arts-and-Sciences departments, the law school, and the medical school. "X State University" frequently started out as "X State College" and only acquired the more august "University" designation within the second half of the twentieth century. It is, more often than not, a land-grant institution, practical-minded, skewed toward agricultural, veterinary, and engineering departments while showing a decent respect for the liberal arts.

Normal Schools—the third tier—were post-secondary institutions whose purpose was to train the teachers who would staff those every-other-mile schoolrooms on the Cartesian road grid. The same inflationary pressure that turned X State College into X State University eventually caused these to get promoted to "University of [geographical modifier] X" or "[geographical modifier] X University," which is how we got the University of Northern Iowa, Eastern Illinois University, and many others.

The result is a network of public universities, typically situated in small cities (population, say, between twenty and two hundred thousand) and scattered about the upper Midwest at intervals of approximately one tank of gas. Precisely because of their proximity (spang in the middle of their catchment areas); their unprepossessing rank in the academic hierarchy; their practical, down-to-earth emphases; and their athletic teams, which entertain the surrounding areas, which are too sparsely populated to support professional squads, these institutions have escaped the censure/taint of elitism or ivory towerism that, deservedly or not, tends to get slapped onto private, coastal universities by those elements of society who, when depicted cinematically, are generally shown brandishing torches and pitchforks. This may have changed during the twenty-first century because of the politicization of science, but none of that existed in the MACT of the mid- to late twentieth century, when most people's attitudes toward science were shaped more by antibiotics, the polio vaccine, and moon rockets than current this-can't-be-happening controversies over evolution and global warming.

According to numerical metrics of selectivity, academic prestige, etc.—and believe me, these are exactly the kinds of yardsticks by which these people rule everything—these schools tend to be somewhere behind the prestigious and older private schools of the coasts (not because the people are any dumber but because it is part of their mission to pull in the whole spectrum of academic talent whereas coastal institutions are lodged in well-defined strata). That, combined with the habitually dour and self-deprecating, not to say passive-aggressive, character of residents of the upper Midwest, has left them with chips on their shoulders and an embarrassing tendency to denote themselves as "the Harvard of the Midwest" or what have you. Seen in a longer perspective and without the overlay of coast-vs.-midwest politics, however, the achievements of the state universities are more remarkable, and certainly more unusual, in that one would not necessarily expect newish, publicly funded institutions to be able to make such respectable showings in competition with far older, privately funded schools that have nothing to do except pile up their endowments century after century and educate the cleverest, best-prepared scions of powerful families.

I describe here a situation that existed during the second

half of the twentieth century. It might be different now. But in those days, graduate students and faculty members U-Hauled from MACT to MACT somewhat in the manner of Arabs oasis-hopping across otherwise inhospitable terrain, and all of the MACTs, *mutatis mutandis*, were the same. Only the school colors and mascots really differed.

Geographical isolation is key to MACT culture. If you have an academic position in, say, greater Boston, you are spending your working days in a culture similar to that of the MACT, but when you go back to your house in Saugus or your apartment in Allston-Brighton, you're in a place where, even if you're not making more money than the people around you, you do enjoy an at least theoretically exalted status by virtue of your advanced degree and your prestigious job. Some people will treat you with a degree of deference. Even those who don't remind you of what an odd duck you are in the larger scheme. Whereas if you are in a MACT you are accorded no sense of specialness whatsoever.

And, remember, these are the professors themselves I'm talking about. The professors' *kids*, growing up in a community where all of the other kids had PhD parents, never acquired in the first place, and so did not have to lose, their sense of belonging to a special, or even an unusual, class.

There are certain other peculiarities of the MACT that might find their place in a longer treatment of the topic, such as the way that garbage collectors' sons and farmers' daughters ended up being treated the same as everyone else, as long as they were smart, and the way that grad students from what were in those days seen as extremely exotic and remote places (Thailand, Afghanistan, Nigeria) were surprised, not always happily, to see their children fully and unquestioningly integrated into small-town Midwestern society, going to keggers and t.p.-ing their friends' houses as if their ancestors had come over on the *Mayflower.*

The premise of this introduction, which will be nailed to the mast very shortly, is that in *Everything and More* David Foster Wallace is speaking in a language and employing a style of inquiry that might strike people who have not breathed the air of Ames, Bloomington-Normal, and Champaign-Urbana as unusual enough to demand some sort of an explanation. And that, lacking such background, many of DFW's critics fall into a common pattern of error, which consists of attempting to explain his style and approach by imputing certain stances or motives to him, then becoming nonplussed, huffy, or downright offended by same. It's a mistake that befuddles MACT natives who see this book as simply what it is: one of the other smart kids trying to explain some cool stuff.

The regrettable fact that (barring possible random playground encounters) I never actually met Mr. Wallace is not necessarily a disqualification from writing an introduction. For that, all that is strictly required is some familiarity with the work being introduced. But since anyone can read *Everything and More,* that hardly makes for a unique, or even an unusual, qualification, and so my strategy here will be to predicate certain things of DFW and his work, based solely on our common MACT provenance, that are wild guesses but that I'm pretty sure are right. This could be developed at heinous length, but since what you are reading is merely an introduction to the actual book ("booklet"—DFW) I am going to lay my core thesis directly on the line and put it to you that this is all about a quintessentially MACTish denial, or at least shrugging-off, of an attitude toward knowledge that in the Greek tradition is conveyed in the story of Prometheus and, in the Judeo-Christian, in that of Eve.

Here, in a conjectural version of this introduction that was more dignified and old-school, those two myths would be recounted and glossed. Matters being what they are, I will encourage anyone unfamiliar with them to consult Google before proceeding. These are meant to be scary, cautionary tales to keep Bronze Age peons from asking difficult questions of their betters. To say that they have outlived their usefulness is wrong, since they were never useful to begin with. At some level, though, we've all imbibed them and they can be invoked in rhetoric to elicit certain predictable responses. By and large, these enure to the benefit of those who have acquired lots of knowledge. You might not think so, for the Promethean myth is ostensibly a knock on academics. Not so ostensibly, though, it gives scientists a reason to put on priestly airs and, by hinting at the perhaps not-so-priestly stances of their counterparts in other countries, haul down defense grants. And it gives non-scientists an implicit pitchfork to brandish in the scientists' faces. Accordingly, a kind of deal has been struck in which both scientists and non-scientists have ended up accepting the Promethean myth as being a passable model of reality. Call this the Promethean consensus. The Promethean consensus is something that no one would ever admit to believing in, if you pinned them down and tried to get them to engage in that level of introspection, but is universally hammered home by every movie and television show about science and a good many books as well, and obviously underlies the public postures that scientists are expected to adopt.

Once you've bought into it, the only two stances you can really take toward the Promethean consensus are to respect its rules or to wilfully break them. You are either a priest or a bad boy. Priest because, if you are one of the keepers of the academic flame and are willing to allow that some of your knowledge is dangerous, you can get a lot of mileage out of intoning the right solemn and portentous sound bites. Bad boy because the downside of the Promethean myth has largely gone away. No one is getting expelled from the Garden of Eden or being chained to a rock to have his liver torn by vultures anymore. It's true that modern-day scientists have to take their share of flak, but, with the exception of people who run girls' schools in Afghanistan, or the occasional biomed researcher who's run afoul of animal-rights activists, they no longer have to dodge pitchforks. And so if you're one of the people who actually has access to Promethean-grade knowledge, there's no longer much personal risk, and, to the extent that the knowledge is perceived as dangerous, it can just feel kind of cool, in a naughty way, like you're a teenager who just figured out where Dad hides the keys to his gun cabinet.

Neither of these seemed to be going on with the irradiated corn seeds. Clearly, giving that kind of stuff to kids is non-priestly behavior. But when they were handed out at the scout meeting, or when we were exposed to sacred knowledge in countless other ways in the MACT, it was never done with an attitude of "we're getting away with something—aren't we being naughty" but rather "here's some interesting and perhaps useful knowledge that any well-brought-up young person will want to have."

So the Promethean consensus is not much in evidence in the MACT. After I went coastal, I committed a string of social gaffes in which I failed to address or introduce some PhD-endowed person with the correct title. We simply never did this where I grew up because it would have given us the faintly comical effect of characters in *The Crucible* addressing one another as "Goodman this" and "Goodwife that" (in our town there was one man, not employed in academia, who had a PhD, and who insisted on being addressed by his title. The view taken of him by everyone else might most politely be described as bemused).

In the preceding paragraph I am using a somewhat tawdry rhetorical shortcut by making fun of people who are pompous about academic titles, and readers from academic, but non-MACT, environments are probably getting hot under the collar and feeling as though they've been ill-used by a thoroughly odious hit-and-run straw-man argument, so let me make clear right away that it's way more complicated than I'm making it sound, and that professors at Harvard and Cambridge and Bologna and Berkeley address one another by their first names all the time.

I am, however crudely, trying to direct the reader's attention to the fact that, even among academics who ride bicycles to work and wear T-shirts and blue jeans and eschew use of formal titles, there are certain strictures and rules and bright lines and hierarchies that Must Be Respected and that people who violate them can find themselves the object of crazily vehement retribution. And here I feel I am on firmer rhetorical ground, since anyone who has spent time on any rung of that ladder will probably have at least one face-burning anecdote about how he or she ran afoul of these strictures and got crucified in a faculty meeting or a letter to the editor or a rampant email thread. I put it to you that, improbable as it might seem, MACT natives can grow up not being keenly aware of those rules, somewhat as the Eloi never twigged to the fact that they were Morlock chow. As I have tried to demonstrate with the irradiated-corn anecdote, the MACT breeds an anti-Promethean nonchalance that really rubs some people the wrong way. Every paragraph of *Everything and More* is imbued with it.

It is an expectation, and a reasonable enough one, that anyone who ventures to write about mathematics must make some kind of positive advance or else shut up. Exceptions are made for occasional review articles, which summarize other results without presenting new material per se, but even a review needs to be written to sufficiently exacting standards that a serious, let us say PhD-level, student of the field in question can take every statement in the thing at face value and never be exposed to the risk that some part of it, in retrospect, might be found to have been glossed over, rearranged, or out-and-out screwed up. So if one is playing by the rules of academic publishing, writing an intellectually serious book about math that engages in some rearrangement and glossing over, as DFW explicitly does in *Everything and More*, is not looked on favorably.

Another practice that seems to make tenured academics practically hop up and down in rage is the crossing of boundaries between sub-sub-disciplines (or, in the case of history, geographical regions or chronological epochs) to write articles that pull together a number of threads and point out common themes among them. The exact reasons for this taboo are probably best left to anthropologists or psychologists, but I infer that this sort of thing is viewed as a privilege gained only with age and emeritus-level distinction and that to write any such material before the age of sixty gets one designated as a whippersnapper, which, in the academic world, is the setup for retributive measures of a severity normally seen only in Greek myths.

So the rules of the academic publishing road are both strict and cruelly enforced. This imposes some narrow and hard limits on what smart people can get away with writing about, which are sufficiently restrictive that some effort goes into finding loopholes. The biggest of these appears to be science fiction. SF novelists arrogate to themselves and, by convention, are readily afforded, a kind of court jester's immunity. And indeed there have been any number of hard science professors who have donned the motley, taken up the pen, and written more or less successful works of hard science fiction as a way of dodging those two terrible strictures against popularization/simplification and synoptic pulling-together-of-diverse-strands.

It is also permissible for serious academics to write books that are explicitly targeted at general readers, though again this tends to be viewed as whippersnapperish behavior if indulged in too early in one's career.

To this point, then, we have two categories of books-about-real-science-for-non-specialist-readers: the hard SF novel and the popularizing book written by an actual scientist. There is a third category, in which a writer, well-educated but without formal credentials in the field in question, immerses himself in the subject matter and then does his level best to explain it. There is a tendency, which is by no means a bad thing, for such books to become somewhat self-referential and autobiographical as the author tells the tale of his own self-education. While the premise, explained this way, sounds dodgy, these books can be really good, since the writer knows what it's like to not understand the material, and can tell the story of learning it as a narrative.

A fourth category, seemingly quite different from Type 3 but in some ways similar, is the History of Science book, which generally takes the form of a narrative about the efforts of one or more scientists to figure something out. Here the questing author of the Type 3 book is replaced, as protagonist, by the actual scientist who figured it all out in the first place.

Again, this introduction might be a more respectable—certainly it would be longer—document if it now listed specific examples of each of the above-mentioned four types of books and engaged in some actual literary criticism. But anyone who is bothering to read an introduction by an SF novelist to a book about infinity by DFW probably has examples of all four types on her bookshelf and so this will be left, as the saying goes, as an exercise for the reader. Just to be clear, though, I will list some examples:

Type 1: Any fiction by Gregory Benford
Type 2: *A Brief History of Time* by Stephen Hawking
Type 3: *1491* by Charles Mann
Type 4: *Einstein in Berlin* by Tom Levenson

What is clearly true about all of these types of books is that they are safe to write, in the sense that critically minded readers from the academic world will fairly quickly say to themselves, "ah, this is one of those" and then, if they wish to criticize them, will do so according to the rules of that type.

Everything and More occupies a hard-to-pin-down space in the Venn diagram that has been taking shape in preceding paragraphs (and before going into detail on that, I'll just supply the premonitory information that books without a clear coordinate on the Venn diagram tend to make people crazy, since this makes it unclear which set of interpretive and critical ground rules is to be applied).

To begin with, DFW was arguably a science fiction writer (*Infinite Jest*), although he probably would not have classified himself as such. Of course *Everything and More* is not SF, or even F, at all, *pace* some of its detractors, but the mere fact of DFW's having been an SF kind of guy muddies the taxonomic waters before we have even gotten started. Novelists—who almost by definition hold motley and informal credentials, when they are credentialed at all—make for an uneasy fit with the academic world, where credentials are everything. And writers who produce books on technical subjects aimed at non-technical readers are doomed to get cranky reviews from both sides: anything short of a fully peer-reviewed monograph is simply wrong and subject to censure from people whose job it is to get it right, and any material that requires unusual effort to read undercuts the work's claim to be accessible to a general audience. So in writing a book such as *Everything and More*, DFW reminds us of the soldier who earns a medal by calling in an artillery strike on his own position, with the possible elaboration that in this case he's out in the middle of no-man's land calling in strikes from both directions.

DFW's degree was in modal logic, which, if you haven't seen

it, is indistinguishable by almost all laymen from pure math, though even more punishingly abstract than mathematics could ever be. Though he did not pursue that career to a PhD and an academic post, the fact that he was able to study such a recondite field at all clearly marks him out as having had what it took to be a hard science/math/logic professional and, thus, in the eyes of hard-math critics, as fair game. We must therefore ask whether *Everything and More* is to be taken as a serious technical book by an actual scientist, or a popularization. Its editors clearly asked for the latter and eventually took delivery of something closer to the former. Which is not to say that DFW makes actual technical advances in mathematics—he doesn't, and doesn't try or claim to—but that he immersed himself in the material in a way that the editors of this series could not reasonably have asked or expected any writer to do, and pitched many parts of the text at a higher technical level than is generally considered a good move in books whose mission it is to popularize science. Which, if all DFW cared about was getting a uniformly rapturous critical reception, might not have been the best tactical approach. But he doesn't appear to have been this kind of guy at all.

In immune-system lingo, the equation-laden sections of *Everything and More* cause it to express certain antigens that arouse the retributive ardor of hard-science and math reviewers. The analogy being apposite here because the immune system, when aroused, can elicit a range of reactions from a mild sense that something isn't right, to irritation, to hives, to full-on T-cell counterattack and organ rejection.

Finally, *Everything and More*, in many sections, alternates between being a Type 3 and a Type 4 (see above taxonomic breakdown) in that, part of the time, we are getting autobiographical material about how DFW learned mathematics, mostly under one Dr. Goris, and part of the time it becomes a History of Science book in which we learn about the lives and careers of Dedekind, Weierstrass, Cantor, et al.

Having as it were set all of those pieces out on the board, the weakest possible claim that I can now assert is that I really like this book and that, as I was reading it, it never even occurred to me to be troubled, confused, annoyed, or nonplussed by any of the features alluded to: the fact that it was written by a fiction writer, the excursions into highly technical discourse, the caveats—clearly and repeatedly stated by DFW—that the technical bits simplified and glossed over material in a way that wouldn't be satisfactory to mathematicians, and the use of both autobiographical, and just plain biographical, material. My advice, therefore, dear reader, is that you simply read it, and that if you happen to be a math major, you then peruse some of the trenchant criticisms of the book that have appeared in the mathematical literature, improve your understanding of the pure-math content by studying peer-reviewed documents on the same topics, and, in general, make sure that this is not the last thing you read on the topic before your orals.

Having supplied that exhortation, I will add one piece of advice about how to read this book, which is to relax and pay no attention—beyond, of course, reading and enjoying it—to one feature of this book that has engendered an absurd volume of critical boggling, namely, DFW's habit of employing informal pop/slang expressions in close juxtaposition with high-end vocabulary and while talking about fancy stuff. This is nothing except good writing. The vernacular is often the most expressive wing of the language. DFW could write high-powered prose better than just about anyone, but he well knew the value of mixing it with informal day-to-day English, and, though he was especially good at it, it's worth keeping in mind that he was hardly the first great English writer to do so. For every Milton who kept it all on an elevated plane, there was a Shakespeare who knew how to sock us in the chops with some well-timed plain talk (among reviewers with humanities degrees, it also seems compulsory to make some remark—or, just as well, to go on at some length—on "post-modernism," a topic of zero interest to most actual readers).

I infer that some whose academic reputations have been put into play by the assignment to write a review of this book have felt provoked or confused by DFW's disinclination or outright refusal to don the mortarboard—the lofty academic style of expression—that's expected of people who want to thrive within that system but that can be swapped out, by novelists, in favor of the court jester's cap 'n' bells. A dead giveaway being the habit of following a quote from DFW's prose by "[sic]." As long as you are not the sort of person who is in the habit of using "[sic]" after quoting others' work in your own written communications, you should be okay with the style in which *Everything and More* has been written.

The foregoing has been all negative, not in the pop-psych sense of adopting a dispiriting tone, but in the purely technical sense that it has been about negating a number of predicates (DFW didn't buy into the Promethean consensus, *Everything and More* doesn't fit into such-and-such bubble on the Venn diagram, certain criticisms of the book aren't that interesting or useful to most readers). I would like to end with something positive (in both the pop-psych and the technical senses). DFW's writing reflects an attitude that is lovely: a touching, and for the most part well-founded, belief that you can explain anything with words if you work hard enough and show your readers sufficient respect. While it has probably existed in other times and places, it is a Midwestern American College Town attitude all the way.

As an explanation for milder allergic reactions—and, having proselytized DFW's writing to many friends over the years, I've seen a few—some readers posit (often vaguely and fretfully) that there is some archness or smart-assery in DFW's literary style. This, to me anyway, is an unsupportable conclusion, given the obvious love that DFW brings to what he's writing about and his explicitly stated opposition to irony-as-lifestyle in his essay "E Unibus Pluram." Why do people see it when it's not there? It's something to do with the fact that his conspicuous verbal talent and wordplay create a nagging sense among some readers that there's a joke here that they're not getting or that they are somehow being made fools of by an agile knave. Which DFW was not.

To me *Everything and More* reads, rather, as a discourse from a green, gridded prairie heaven, where irony-free people who've been educated to a turn in those prairie schoolhouses and great-but-unpretentious universities sit around their dinner tables buttering sweet corn, drinking iced tea, and patiently trying to explain even the most recondite mysteries of the universe, out of a conviction that the world must be amenable to human understanding and that if you can understand something, you can explain it in words: fancy words if that helps, plain words if possible. But in any case you can reach out to other minds through that medium of words and make a connection. Handing out irradiated corn kernels to a troop of Boy Scouts and writing books that explain difficult matters in disarmingly informal language are the same act, a way of saying "here is something cool that I want to share with you for no reason other than making the spark jump between minds." If that is how you have been raised, then to explain anything to anyone is a pleasure. To explain difficult things is a challenge. And to explain the infamously difficult ideas that were spawned in chiliastic profusion during the late nineteenth and early twentieth centuries (Infinities, Relativity, Quantum Mechanics, Hilbert's problems, Gödel's Proof) is Mount Everest.

So in reading *Everything and More*, cleverness or verbal pyrotechnics or archness is not the emotional tone that comes through to me, but a kind of open-soulness and desire to connect that were touching before, and heartbreaking after, David Foster Wallace succumbed, at the age of forty-six, to a cruel and incurable disease. Because of this we will not have the opportunity to enjoy and profit from many other explanations that it was in his power to supply on diverse topics, lofty and mundane, and so we must content ourselves with what he did leave behind—an impossibility given the pleasure and the insight he gave us in *Everything and More,* and his obvious ability to have provided much more, had fortune treated him with as much consideration as he did his readers.

Small But Necessary Foreword

Unfortunately this is a Foreword you actually have to read—and first—in order to understand certain structural idiosyncrasies and bits of what almost look like code in the main text. Of the latter the most frequent is a boldface 'IYI'. This, be apprised, is not a tic or typo but instead stands for the clause *If you're interested*, which was getting used over and over so many times in early drafts that what eventually happened is that through sheer repetition it evolved from a natural-language phrase for introducing some clause into an abstract extratextual sign—**IYI**—that now serves to classify certain chunks of text in a particular way. Which way will now be justified and explained.

Like the other booklets in this 'Great Discoveries' series, *Everything and More* is a piece of pop technical writing. Its subject is a set of mathematical achievements that are extremely abstract and technical, but also extremely profound and interesting, and beautiful. The aim is to discuss these achievements in such a way that they're vivid and comprehensible to readers who do not have pro-grade technical backgrounds and expertise. To make the math beautiful—or at least to get the reader to see how someone might find it so. Which of course all sounds very nice, except there's a hitch: just how technical can the presentation get without either losing the reader or burying her in endless little definitions and explanatory asides? Plus if you assume, as seems plausible, that some readers are going to have much stronger tech backgrounds than others, how can the discussion be pitched so that it's accessible to the neophyte without being dull or annoying to somebody who's had a lot of college math?

In the following document, the boldface 'IYI' designates bits of material that can be perused, glanced at, or skipped altogether if the reader wants. Meaning skipped without serious loss. Over half the document's footnotes are probably **IYI**, as well as several different ¶s and even a couple subsections of the main text. Some of the optional bits are digressions or bits of historical ephemera[1]; some are definitions or explanations that a math-savvy reader won't need to waste time on. Most **IYI**-grade chunks, though, are designed for readers with strong technical backgrounds, or unusual interest in actual math, or preternatural patience, or all three; they (the

[1] **IYI** Here's a good example of an **IYI** factoid. Your author here is someone with a medium-strong amateur interest in math and formal systems. He is also someone who disliked and did poorly in every math course he ever took, save one, which wasn't even in college, but which was taught by one of those rare specialists who can make the abstract alive and urgent, and who actually really talks to you when he's lecturing, and of whom anything that's good about this booklet is a pale and well-meant imitation.

chunks) provide a more detailed look at stuff that the main discussion glosses or breezes through.

There are other abbreviations in the booklet, too. Some are just to save space. Others are the consequence of a strange stylistic problem in tech writing, which is that the same words often have to get used over and over in a way that would be terribly clunky in regular prose—the thing is that some technical words have highly specific denotations that no synonym can capture. Which means that, especially respecting certain high-tech proper nouns, abbreviation is the only way to achieve any kind of variation at all. None of that is really your problem. All the booklet's abbreviations are contextualized in such a way that it ought to be totally clear what they stand for; but in case of authorial foul-ups or unnecessary confusion, here is a list of the main ones, which can be flipped back and referred to if necessary:

1-1C	= One-to-One Correspondence
A.C.	= Axiom of Choice
A.S.T.	= Axiomatic Set Theory
ATH	= Fourier's *Analytic Theory of Heat*
B.T.	= Binomial Theorem
B.W.T.	= Bolzano-Weierstrass Theorem
"C. and I.R."	= Dedekind's "Continuity and Irrational Numbers"
C.H.	= Continuum Hypothesis
C.P.	= Cartesian Product
D.B.P.	= Divine Brotherhood of Pythagoras
D.E.	= Differential Equation
D.P.	= Diagonal Proof
E.G.	= EMERGENCY GLOSSARY

E.V.T.	= Weierstrass's Extreme Values Theorem
F.T.C.	= Fundamental Theorem of the Calculus
G.C.P.F.S.	= General Convergence Problem of Fourier Series
L.A.P.	= Limited Abstraction Principle
LEM	= Law of the Excluded Middle
N. & L.	= Newton and Leibniz
N.L.	= Number Line
N.S.T.	= Naïve Set Theory
O.O.M.	= Plato's One Over Many argument
P.I.	= Principle of Induction
P of the I	= Bolzano's *Paradoxes of the Infinite*
P.S.A.	= Power Set Axiom
P.T.	= Pythagorean Theorem
R.L.	= Real Line
TNS	= Galileo's *Two New Sciences*
U.A.P.	= Unlimited Abstraction Principle
U.T.	= Uniqueness Theorem
VC	= Vicious Circle
VIR	= Vicious Infinite Regress
VNB	= Von Neumann–Bernays system of axioms for set theory
V.S.P.	= Vibrating String Problem
W.E.	= Wave Equation
ZFS	= Zermelo–Fraenkel–Skolem system of axioms for set theory
Z.P.	= Zeno's Paradox

§1a. There is such a thing as an historian of mathematics. Here is a nice opening-type quotation from one such historian in the 1930s:

> One conclusion appears to be inescapable: without a consistent theory of the mathematical infinite there is no theory of irrationals; without a theory of irrationals there is no mathematical analysis in any form even remotely resembling what we now have; and finally, without analysis the major part of mathematics—including geometry and most of applied mathematics—as it now exists would cease to exist. The most important task confronting mathematicians would therefore seem to be the construction of a satisfactory theory of the infinite. Cantor attempted this, with what success will be seen later.

The sexy math terms don't matter for now. The Cantor of the last line is Prof. Georg F. L. P. Cantor, b. 1845, a naturalized German of the merchant class and the acknowledged father of abstract set theory and transfinite math. Some historians have argued back and forth about whether he was Jewish. 'Cantor' is just Latin for singer.

G. F. L. P. Cantor is the most important mathematician of the nineteenth century and a figure of great complexity and pathos. He was in and out of mental hospitals for much of his later adulthood and died in a sanitarium in Halle[1] in 1918. K. Gödel, the most important mathematician of the twentieth century, also died as the result of mental illness. L. Boltzmann,

[1] **IYI** Halle, a literal salt mine just upriver from Leipzig, is best known as Handel's hometown.

the most important mathematical physicist of the nineteenth century, committed suicide. And so on. Historians and pop scholars tend to spend a lot of time on Cantor's psychiatric problems and on whether and how they were connected to his work on the mathematics of ∞.

At Paris's 2nd International Congress of Mathematicians in 1900, Prof. D. Hilbert, then the world's #1 mathematician, described Georg Cantor's transfinite numbers as "the finest product of mathematical genius" and "one of the most beautiful realizations of human activity in the domain of the purely intelligible."

Here is a quotation from G. K. Chesterton: "Poets do not go mad; but chess players do. Mathematicians go mad, and cashiers; but creative artists very seldom. I am not attacking logic: I only say that this danger does lie in logic, not in imagination." Here also is a snippet from the flap copy for a recent pop bio of Cantor: "In the late nineteenth century, an extraordinary mathematician languished in an asylum. . . . The closer he came to the answers he sought, the further away they seemed. Eventually it drove him mad, as it had mathematicians before him."

The cases of great mathematicians with mental illness have enormous resonance for modern pop writers and filmmakers. This has to do mostly with the writers'/directors' own prejudices and receptivities, which in turn are functions of what you could call our era's particular archetypal template. It goes without saying that these templates change over time. The Mentally Ill Mathematician seems now in some ways to be what the Knight Errant, Mortified Saint, Tortured Artist, and Mad Scientist have been for other eras: sort of our

Prometheus, the one who goes to forbidden places and returns with gifts we all can use but he alone pays for. That's probably a bit overblown, at least in most cases.[2] But Cantor fits the template better than most. And the reasons for this are a lot more interesting than whatever his problems and symptoms were.[3]

Merely knowing about Cantor's accomplishments is different from appreciating them, which latter is the general project here and involves seeing transfinite math as kind of like a tree, one with its roots in the ancient Greek paradoxes of continuity and incommensurability and its branches entwined in the modern crises over math's foundations—Brouwer and Hilbert and Russell and Frege and Zermelo and Gödel and Cohen et al. The names right now are less important than the tree thing, which is the main sort of overview-trope you'll be asked to keep in mind.

[2] **IYI** although so is the other, antipodal stereotype of mathematicians as nerdy little bowtied fissiparous creatures. In today's archetypology, the two stereotypes seem to play off each other in important ways.

[3] In modern medical terms, it's fairly clear that G. F. L. P. Cantor suffered from manic-depressive illness at a time when nobody knew what this was, and that his polar cycles were aggravated by professional stresses and disappointments, of which Cantor had more than his share. Of course, this makes for less interesting flap copy than Genius Driven Mad By Attempts To Grapple With ∞. The truth, though, is that Cantor's work and its context are so totally interesting and beautiful that there's no need for breathless Prometheusizing of the poor guy's life. The real irony is that the view of ∞ as some forbidden zone or road to insanity—which view was very old and powerful and haunted math for 2000+ years—is precisely what Cantor's own work overturned. Saying that ∞ drove Cantor mad is sort of like mourning St. George's loss to the dragon: it's not only wrong but insulting.

§1b. Chesterton above is wrong in one respect. Or at least imprecise. The danger he's trying to name is not logic. Logic is just a method, and methods can't unhinge people. What Chesterton's really trying to talk about is one of logic's main characteristics—and mathematics'. Abstractness. Abstraction.

It is worth getting straight on the meaning of *abstraction.* It's maybe the single most important word for appreciating Cantor's work and the contexts that made it possible. Grammatically, the root form is the adjectival, from the L. *abstractus* = 'drawn away'. The *O.E.D.* has nine major definitions of the adjective, of which the most apposite is 4.a.: "Withdrawn or separated from matter, from material embodiment, from practice, or from particular examples. Opposed to *concrete.*" Also of interest are the *O.E.D.*'s 4.b., "Ideal, distilled to its essence," and 4.c., "Abstruse."

Here is a quotation from Carl B. Boyer, who is more or less the Gibbon of math history[4]: "But what, after all, are the integers? Everyone thinks that he or she knows, for example, what the number three is—until he or she tries to define or explain it." W/r/t which it is instructive to talk to 1st- and 2nd-grade math teachers and find out how children are actually taught about integers. About what, for example, the number five is. First they are given, say, five oranges. Something they can touch or hold. Are asked to count them. Then they

[4] **IYI** Boyer is joined at the top of the math-history food chain only by Prof. Morris Kline. Boyer's and Kline's major works are respectively *A History of Mathematics* and *Mathematical Thought from Ancient to Modern Times.* Both books are extraordinarily comprehensive and good and will be liberally cribbed from.

are given a picture of five oranges. Then a picture that combines the five oranges with the numeral '5' so they associate the two. Then a picture of just the numeral '5' with the oranges removed. The children are then engaged in verbal exercises in which they start talking about the integer 5 per se, as an object in itself, apart from five oranges. In other words they are systematically fooled, or awakened, into treating numbers as things instead of as symbols for things. Then they can be taught arithmetic, which comprises elementary relations between numbers. (You will note how this parallels the ways we are taught to use language. We learn early on that the noun 'five' means, symbolizes, the integer 5. And so on.)

Sometimes a kid will have trouble, the teachers say. Some children understand that the word 'five' stands for 5, but they keep wanting to know 5 *what?* 5 oranges, 5 pennies, 5 points? These children, who have no problem adding or subtracting oranges or coins, will nevertheless perform poorly on arithmetic tests. They cannot treat 5 as an object per se. They are often then remanded to Special Ed Math, where everything is taught in terms of groups or sets of actual objects rather than as numbers "withdrawn from particular examples."[5]

[5] B. Russell has an interesting ¶ in this regard about high-school math, which is usually the next big jump in abstraction after arithmetic:

> In the beginning of algebra, even the most intelligent child finds, as a rule, very great difficulty. The use of letters is a mystery, which seems to have no purpose except mystification. It is almost impossible, at first, not to think that every letter stands for some particular number, if only the teacher would reveal *what* number it stands for. The fact is, that in algebra the mind is first taught to consider general truths, truths which are not asserted to hold only of this or that particular thing, but of any one of a whole group of things. It is in

The point: The basic def. of 'abstract' for our purposes is going to be the somewhat concatenated 'removed from or transcending concrete particularity, sensuous experience'. Used in just this way, 'abstract' is a term from metaphysics. Implicit in all mathematical theories, in fact, is some sort of metaphysical position. The father of abstraction in mathematics: Pythagoras. The father of abstraction in metaphysics: Plato.

The *O.E.D.*'s other defs. are not irrelevant, though. Not just because modern math is abstract in the sense of being extremely abstruse and arcane and often hard to even look at on the page. Also essential to math is the sense in which abstracting something can mean reducing it to its absolute skeletal essence, as in the abstract of an article or book. As such, it can mean thinking hard about things that for the most part people can't think hard about—because it drives them crazy.

All this is just sort of warming up; the whole thing won't be like this. Here are two more quotations from towering figures. M. Kline: "One of the great Greek contributions to the very concept of mathematics was the conscious recognition and emphasis of the fact that mathematical entities are abstractions, ideas entertained by the mind and sharply distinguished from physical objects or pictures." F.d.l. Saussure: "What has escaped philosophers and logicians is that from the moment a system of symbols becomes independent

---

the power of understanding and discovering such truths that the mastery of the intellect over the whole world of things actual and possible resides; and ability to deal with the general as such is one of the gifts that a mathematical education should bestow.

of the objects designated it is itself subject to undergoing displacements that are incalculable for the logician."

Abstraction has all kinds of problems and headaches built in, we all know. Part of the hazard is how we use nouns. We think of nouns' meanings in terms of denotations. Nouns stand for things—*man, desk, pen, David, head, aspirin.* A special kind of comedy results when there's confusion about what's a real noun, as in 'Who's on first?' or those *Alice in Wonderland* routines—'What can you see on the road?' 'Nothing.' 'What great eyesight! What does nothing look like?' The comedy tends to vanish, though, when the nouns denote abstractions, meaning general concepts divorced from particular instances. Many of these abstraction-nouns come from root verbs. 'Motion' is a noun, and 'existence'; we use words like this all the time. The confusion comes when we try to consider what exactly they mean. It's like Boyer's point about integers. What exactly do 'motion' and 'existence' denote? We know that concrete particular things exist, and that sometimes they move. Does motion per se exist? In what way? In what way do abstractions exist?

Of course, that last question is itself very abstract. Now you can probably feel the headache starting. There's a special sort of unease or impatience with stuff like this. Like 'What exactly is existence?' or 'What exactly do we mean when we talk about motion?' The unease is very distinctive and sets in only at a certain level in the abstraction process—because abstraction proceeds in levels, rather like exponents or dimensions. Let's say 'man' meaning some particular man is Level One. 'Man' meaning the species is Level Two. Something like 'humanity' or 'humanness' is Level Three; now we're talking about the abstract criteria for something qualifying as human.

And so forth. Thinking this way can be dangerous, weird. Thinking abstractly enough about anything . . . surely we've all had the experience of thinking about a word—'pen,' say—and of sort of saying the word over and over to ourselves until it ceases to denote; the very strangeness of calling something a pen begins to obtrude on the consciousness in a creepy way, like an epileptic aura.

As you probably know, much of what we now call analytic philosophy is concerned with Level Three– or even Four–grade questions like this. As in epistemology = 'What exactly is knowledge?'; metaphysics = 'What exactly are the relations between mental constructs and real-world objects?'; etc.[6] It might be that philosophers and mathematicians, who spend a lot of time thinking (a) abstractly or (b) about abstractions or (c) both, are *eo ipso* rendered prone to mental illness. Or it might just be that people who are susceptible to mental illness are more prone to think about these sorts of things. It's a chicken-and-egg question. One thing is certain, though. It is a total myth that man is by nature curious and truth-hungry and wants, above all things, *to know.*[7] Given certain recognized senses of 'to know,' there is in fact a great deal of stuff we do *not* want to know. Evidence for this is the enormous number of very basic questions and issues we do not like to think about abstractly.

[6] **IYI** According to most sources, G. F. L. P. Cantor was not just a mathematician—he had an actual Philosophy of the Infinite. It was weird and quasi-religious and, not surprisingly, abstract. At one point Cantor tried to switch his U. Halle job from the math dept. to philosophy; the request was turned down. Admittedly, this was not one of his stabler periods.

[7] **IYI** The source of this pernicious myth is Aristotle, who is in certain respects the villain of our whole Story—q.v. §2 sub.

Theory: The dreads and dangers of abstract thinking are a big reason why we now all like to stay so busy and bombarded with stimuli all the time. Abstract thinking tends most often to strike during moments of quiet repose. As in for example the early morning, especially if you wake up slightly before your alarm goes off, when it can suddenly and for no reason occur to you that you've been getting out of bed every morning without the slightest doubt that the floor would support you. Lying there now considering the matter, it appears at least theoretically possible that some flaw in the floor's construction or its molecular integrity could make it buckle, or that even some aberrant bit of quantum flux or something could cause you to melt right through. Meaning it doesn't seem logically impossible or anything. It's not like you're actually scared that the floor might give way in a moment when you really do get out of bed. It's just that certain moods and lines of thinking are more abstract, not just focused on whatever needs or obligations you're going to get out of bed to attend to. This is just an example. The abstract question you're lying there considering is whether you are truly justified in your confidence about the floor. The initial answer, which is yes, lies in the fact that you've gotten out of bed in the morning thousands—actually well over ten thousand times so far, and each time the floor has supported you. It's the same way you're also justified in believing that the sun will come up, that your wife will know your name, that when you feel a certain set of sensations it means you're getting ready to sneeze, & c. Because they've happened over and over before. The principle involved is really the only way we can predict any of the phenomena we just automatically count on without having to think about them. And the vast bulk of daily life is composed of these sorts of phenomena; and without this confidence based on past experience we'd all go insane, or at least we'd be unable to function because we'd have to stop and deliberate about every last little thing. It's a fact: life as we know it would be impossible without this confidence. Still, though: Is the confidence actually justified, or just highly convenient? This is abstract thinking, with its distinctive staircase-shaped graph, and you're now several levels up. You're no longer thinking just about the floor and your weight, or about your confidence re same and how necessary to basic survival this kind of confidence seems to be. You're now thinking about some more general rule, law, or principle by which this unconsidered confidence in all its myriad forms and intensities is in fact justified instead of being just a series of weird clonic jerks or reflexes that propel you through the day. Another sure sign it's abstract thinking: You haven't moved yet. It feels like tremendous energy and effort is being expended and you're still lying perfectly still. All this is just going on in your mind. It's extremely weird; no wonder most people don't like it. It suddenly makes sense why the insane are so often represented as grabbing their head or beating it against something. If you had the right classes in school, however, you might now recall that the rule or principle you want does exist—its official name is the Principle of Induction. It is the fundamental precept of modern science. Without the Principle of Induction, experiments couldn't confirm a hypothesis, and nothing in the physical universe could be predicted with any confidence at all. There could be no natural laws or scientific truths. The P.I. states that if something *x* has happened in certain particular circumstances *n* times in the past, we are justified in believing that the same circumstances will produce x on the $(n + 1)$th occasion. The P.I. is wholly respectable and authoritative, and it seems like a well-lit exit out of the whole problem. Until, that is, it happens to strike you (as can occur only in very abstract moods or when there's an unusual amount of time before the alarm goes off) that the P.I. is *itself* merely an abstraction from experience . . . and so now what exactly is it that justifies our confidence in the P.I.? This latest thought may or may not be accompanied by a concrete memory of several weeks spent on a relative's farm in childhood (long story). There were four chickens in a wire coop off the garage, the brightest of whom was called Mr. Chicken. Every morning, the farm's hired man's appearance in the coop area with a certain burlap sack caused Mr. Chicken to get excited and start doing warmup-pecks at the ground, because he knew it was feeding time. It was always around the same time t every morning, and Mr. Chicken had figured out that $t(\text{man} + \text{sack}) = \text{food}$, and thus was confidently doing his warmup-pecks on that last Sunday morning when the hired man suddenly reached out and grabbed Mr. Chicken and in one smooth motion wrung his neck and put him in the burlap sack and bore him off to the kitchen. Memories like this tend to remain quite vivid, if you have any. But with the thrust, lying here, being that Mr. Chicken appears now actually to have been correct—according to the Principle of Induction—in expecting nothing but breakfast from that $(n + 1)$th appearance of man + sack at t. Something about the fact that Mr. Chicken not only didn't suspect a thing but appears to have been *wholly justified* in not suspecting a thing—this seems concretely creepy and upsetting. Finding some higher-level justification for your confidence in the P.I. seems much more urgent when you realize that, without this justification, our own situation is basically indistinguishable from that of Mr. Chicken. But the conclusion, abstract as it is, seems inescapable: what justifies our confidence in the Principle of Induction is that it has always worked so well in the past, at least up to now. Meaning that our only real justification for the Principle of Induction is the Principle of Induction, which seems shaky and question-begging in the extreme.

The only way out of the potentially bedridden-for-life paralysis of this last conclusion is to pursue further abstract side-inquiries into what exactly 'justification' means and whether it's true that the only valid justifications for certain beliefs and principles are rational and noncircular. For instance, we know that in a certain number of cases every year cars suddenly veer across the centerline into oncoming traffic and crash head-on into people who were driving along not expecting to get killed; and thus we also know, on some level, that whatever confidence lets us drive on two-way roads is not 100% rationally justified by the laws of statistical probability. And yet 'rational justification' might not apply here. It might be more the fact that, if you cannot believe your car won't suddenly get crashed into out of nowhere, you just can't drive, and thus that your need/desire to be able to drive functions as a kind of 'justification' of your confidence.[8]

[8] A compelling parallel here is the fact that most of us fly despite knowing that a definite percentage of commercial airliners crash every year. This gets into the various different kinds of knowing v. 'knowing,' though (see §1c below). Plus it involves etiquette, since commercial air travel is public and a kind of group confidence comes into play. This is why turning to inform your seatmate of the precise statistical probability of your

It would be better not to then start analyzing the various putative 'justifications' for your need/desire to be able to drive a car—at some point you realize that the process of abstract justification can, at least in principle, go on forever. The ability to halt a line of abstract thinking once you see it has no end is part of what usually distinguishes sane, functional people—people who when the alarm finally goes off can hit the floor without trepidation and plunge into the concrete business of the real workaday world—from the unhinged.

INTERPOLATION

The tactical reason for sometimes using '∞' instead of 'infinity' in the natural-language text here is that the double-blink strangeness of '∞' serves as a reminder that it's not clear what we're even talking about. Not yet. For instance, beware of thinking that ∞ is just an incredibly, unbelievably enormous number. There are, of course, many such numbers, especially in physics and astronomy—like, if in physics an ultranano-instant of 5×10^{-44} seconds is generally acknowledged to be the smallest time-interval in which the normal concept of continuous time applies (which it is), astronomical data indicates that there have been roughly 6×10^{60} such ultranano-instants since the Big Bang. That's a 6 followed by 60 zeroes. We've all heard about numbers like this, which we usually imagine can

---

plane crashing is not false but cruel: you are messing with the delicate psychological infrastructure of her justification for flying.

IYI Depending on mood/time, it might strike you as interesting that people who cannot summon this strange faith in principles that cannot be rationally justified, and so cannot fly, are commonly referred to as having an 'irrational fear' of flying.

be conceived and manipulated only with really advanced super-cooled computers or something. Actually, there are plenty of numbers too big for any real or even theoretical computer to process. *Bremermann's Limit* is the operative term here. Given limits imposed by basic quantum theory, one H. Bremermann proved in 1962 that "No data processing system, whether artificial or living, can process more than 2×10^{47} bits per second per gram of its mass," which means that a hypothetical supercomputer the size of the earth ($=$ c. 6×10^{27} grams) grinding away for as long as the earth has existed ($=$ about 10^{10} years, with c. 3.14×10^{7} seconds/year) can have processed at most 2.56×20^{92} bits, which number is known as Bremermann's Limit. Calculations involving numbers larger than 2.56×20^{92} are called *transcomputational problems*, meaning they're not even theoretically doable; and there are plenty of such problems in statistical physics, complexity theory, fractals, etc. All this is sexy but not quite germane. What's germane is: Take some such transcomputational number, imagine it's a grain of sand, conceive of a whole beach, or desert, or planet, or even galaxy filled with such sand, and not only will the corresponding 10^{x} number be $<\infty$, but its square will be $<\infty$, and $10^{(x^{(10^{x})})}$ will be $<\infty$, and so on; and actually it's not even right to compare 10^{x} and ∞ arithmetically this way because they're not even in the same mathematical area code—even, as it were, the same dimension. And yet it's also true that some ∞s are bigger than others, as in arithmetically bigger. All this will get discussed; the thing for now is that only after R. Dedekind and G. Cantor is it even possible to talk about infinite quantities and their arithmetic coherently, meaningfully. Hence the point of using '∞'.

IYI The '∞' symbol itself is technically called the *lemniscate* (apparently from the Greek for 'ribbon') and was introduced to math by John Wallis in his 1655 *Arithmetica infinitorum*, which was one of the important preliminaries for Newton's brand of calculus.[9] Wallis's contemporary Thomas Hobbes, something of a mathematical crank, complained in a review that *Arithmetica infinitorum* was too brutally abstract to even try to read, "a scab of symbols," thereby speaking for generations of undergrads to follow. Other names for the lemniscate include 'the love knot' and 'the Cartesian plane curve that satisfies the equation $(x^2 + y^2)^2 = a^2(x^2 - y^2)$'. If, on the other hand, it's treated trigonometrically and called 'the curve that satisfies the polar equation $r^2 = a \cos 2\theta$,' it is also known as *Bernoulli's Lemniscate.*

END INTERPOLATION

§1c. Apropos the whole business of abstractness and nouns' denotations, there is a syndrome that's either a high-level abstraction or some type of strange nominal mutation. 'Horse' can mean this one horse right here, or it can mean the abstract concept, as in 'Horse = hoofed mammal of family *Equidae*'. Same with the word 'horn'; same with 'forehead'. All these can be abstracted from particulars, but we still know they came from particulars. Except what about a unicorn, which seems to result from the combination of the concepts 'horse,' 'horn,' and 'forehead' and thus has its whole origin in the concatenation of abstractions? Meaning we can

[9] **IYI** As it happens, the only thing that kept Wallis from actually inventing differential calculus in *A.i.* was his ignorance of the Binomial Theorem, which is essential to working with infinitesimals—see esp. §4 below.

conjoin and manipulate abstractions to form entities whose nouns have no particular denotations at all. Here the big problem becomes: In what way can we say a unicorn exists that is fundamentally different, less real, than the way abstractions like humanity or horn or integer exist? Which is once again the question: In what way do abstract entities exist, or do they exist at all except as ideas in human minds—i.e., are they metaphysical fictions? This sort of question can keep you in bed all day too. And it hangs over math from the beginning—what is the ontological status of mathematical entities and relations? Are mathematical realities discovered, or merely created, or somehow both? Here is M. Kline again: "The philosophical doctrines of the Greeks limited mathematics in another way. Throughout the classical period they believed that man does not create the mathematical facts: they preexist. He is limited to ascertaining and recording them."

Plus here is another quotation from D. Hilbert, the great early champion of Cantor's transfinites:

> [T]he infinite is nowhere to be found in reality, no matter what experiences, observations, and knowledge are appealed to. Can thought about things be so much different from things? Can thinking processes be so unlike the actual process of things? In short, can thought be so far removed from reality?

And it's true: there is nothing more abstract than infinity. Meaning at least our fuzzy, intuitive, natural-language concept of ∞. It's sort of the ultimate in drawing away from actual experience. Take the single most ubiquitous and oppressive feature of the concrete world—namely that everything ends, is limited, passes away—and then conceive, abstractly, of something without this feature. Analogies to certain ideas of God are obvious; abstraction from all limitation is one way to account for the religious impulse in secular terms. This is a.k.a. the anthropology of religion: a perfect being can be understood as one devoid of all the imperfections we perceive in ourselves and the world, an omnipotent one as without limitations on his will, etc. The fact that it's a pretty dry and doleful way to talk about religion is neither here nor there; the point is that the exact same sort of explanation can be given for where we got the concept of ∞ and what we ultimately mean by all the forms of the word 'infinite' we toss around. Whether it's actually the right explanation, though, involves what it commits us to. Meaning metaphysically. Do we really want to say that ∞ exists only in the way that unicorns do, that it's all a matter of our manipulating abstractions until the noun 'infinity' has no real referent? What about the set of all integers? Start counting at 1, 2, 3, and so on, and realize that you'll never stop, nor your children when you die, nor theirs, and so on. The integers never stop; there is no end. Does the set of all integers compose a real ∞? Or are the integers themselves not really real but just abstractions; plus what exactly is a set, and are sets real or just conceptual devices, etc.? Or are maybe integers and/or sets only 'mathematically real' as opposed to really real, and what exactly is the difference, and might we want to grant ∞ a certain mathematical reality but not the other kind (assuming there's only one other kind)? And at what point do the questions get so abstract and the distinctions so fine and the cephalalgia so bad that we simply can't handle thinking about any of it anymore?

It is in areas like math and metaphysics that we encounter one of the average human mind's weirdest attributes. This is the ability to conceive of things that we cannot, strictly speaking, conceive of. We can conceive in some rough way of what omnipotence is, for instance. At least we can use the word 'omnipotence' with a fair degree of confidence that we know what we're talking about. And yet even a schoolboy's antinomy like 'Can an omnipotent being make something too heavy for him to lift?' points up serious faultlines in our everyday understanding of omnipotence. So there is one more kind of abstraction that's relevant here. This one is more psychological, and very modern.

Obvious fact: Never before have there been so many gaping chasms between what the world seems to be and what science tells us it is. 'Us' meaning laymen. It's like a million Copernican Revolutions all happening at the same time. As in for instance we 'know,' as high-school graduates and readers of *Newsweek*, that time is relative, that quantum particles can be both there and not, that space is curved, that colors do not inhere in objects themselves, that astronomic singularities have infinite density, that our love for our children is evolutionarily preprogrammed, that there is a blind spot in the center of our vision that our brains automatically fill in. That our thoughts and feelings are really just chemical transfers in 2.8 pounds of electrified pâté. That we are mostly water, and water is mostly hydrogen, and hydrogen is flammable, and yet we are not flammable. We 'know' a near-infinity of truths that contradict our immediate commonsense experience of the world. And yet we have to live and function in the world. So we abstract, compartmentalize: there's stuff we know and stuff we 'know'. I 'know' my love for my child is a function of natural selection, but I know I love him, and I feel and act on what I know. Viewed objectively, the whole thing is deeply schizoid; yet the fact of the matter is that as subjective laymen we don't often feel the conflict. Because of course our lives are 99.9% concretely operational, and we operate concretely on what we know, not on what we 'know'.

Again, we're talking about laymen like you and me, not about the giants of philosophy and math, many of whom had famous trouble navigating the real world. Einstein leaving home in his pajamas, Gödel unable to feed himself, and so on. To appreciate what the inner lives of great scientists/mathematicians/metaphysicians are like, we need only lie here and try to form a truly rigorous and coherent idea—as opposed to a fuzzy or *Newsweek*ish idea—of what we *really mean* by 'omnipotent,' or 'integer,' or 'illimitable,' or 'finite but unbounded'. To try to do some disciplined or directed abstract thinking.[10] There's a very definite but inarticulable fuguelike strain involved in this kind of thinking, a sensation that the epilepsis of saying 'pen, pen' over and over is but a faint pale shadow of. One of the quickest routes to this feeling is (from personal A.M. experience) to try to think hard about dimension. There is something I 'know,' which is that spatial dimensions beyond the Big 3 exist. I can even construct a tesseract or hypercube out of cardboard. A weird sort of cube-within-a-cube, a tesseract is a 3D projection of a 4D object in the same way that '⧉' is a 2D projection of a 3D object. The trick is imagining the tesseract's relevant lines and

[10] The unique and redoubtable Dr. E. Robert Goris of U—— Sr. High School's AP Math I and II used to refer to this also as 'private-sector thinking,' meaning actual productive results were expected.

planes at 90° to each other (it's the same with '⧉' and a real cube), because the 4th spatial dimension is one that somehow exists at perfect right angles to the length, width, and depth of our regular visual field. I 'know' all this, just as you probably do . . . but now try to really picture it. Concretely. You can feel, almost immediately, a strain at the very root of yourself, the first popped threads of a mind starting to give at the seams.

W/r/t 'knowing' v. really actually *knowing*, the second kind is what Descartes meant by "clear and distinct apprehension" and what modern slang connotes via 'handle' or 'deal with'. Thus again the epistoschizoid state of the modern lay mind: We feel like we 'know' things that our minds' conceptual apparatus can't really deal with. These are often objects and concepts at the very farthest reaches of abstraction, things we literally cannot imagine: $n > 3$ manifolds, quantum choreography, fractal sets, dark matter, square roots of negatives, Klein Bottles and Freemish Crates and Penrose Stairways. And ∞. Often, these sorts of things are characterized as existing only 'intellectually' or 'mathematically'. It is, again, far from clear what this means, although the terms themselves are child's play to use.

Note, please, that this lay ability to split our awareness and to 'know' things we cannot handle is distinctively modern. The ancient Greeks, for instance, could not do this. Or wouldn't. They needed things neat, and felt you couldn't know something unless you really understood it.[11] It is not an accident that their mathematics included neither 0 nor ∞. Their word for infinity also meant 'mess'.

[11] **IYI** This is why most of Plato (and nearly all of Aristotle) is about trying to conceptualize and systematize the abstract.

The Greek spirit has informed the philosophy and practice of mathematics from the beginning. Mathematical truths are established by logical proof and are extremely neat and clean. It is just this that exempts math from labyrinthine problems like how exactly to justify the Principle of Induction: mathematical relations and proofs are not inductive but deductive, formal. Math is, in other words, a formal system, with 'formal' meaning pure form, 100% abstract. The core idea is that mathematical truths are certain and universal precisely *because* they have nothing to do with the world. If that's a bit opaque, here is a passage from G. H. Hardy's *A Mathematician's Apology*, the most lucid English prose work ever on math:

> "The certainty of mathematics," says [A. N.] Whitehead, "depends on its complete abstract generality." When we assert that 2 + 3 = 5, we are asserting a relation between three groups of 'things'; and these 'things' are not apples or pennies, or things of any one particular sort or another, but *just* things, 'any old things.' The meaning of the statement is entirely independent of the individualities of the members of the groups. All mathematical 'objects' or 'entities' or 'relations,' such as '2,' '3,' '5,' '+,' or '=,' and all mathematical propositions in which they occur, are completely general in the sense of being completely abstract. Indeed one of Whitehead's words is superfluous, since generality, in this sense, *is* abstractness.

In which quotation please note that 'generality' refers not just to the abstractness of individual terms and referents but to the complete abstract *universality* of the truths asserted. This is the difference between a mere math factoid and a mathematical theorem. A famous example of this difference (famous to students of Dr. Goris, anyway) is that (1) 'The sum of the series $(1 + 3 + 5 + 7 + 9) = 5^2$' is a factoid, whereas (2) 'For any *x*, the sum of the first *x* odd integers = x^2' is a theorem, i.e. actual math.

What follows here is mostly intended as a reminder of stuff you already know in a rough way or had in school. If your familiarity with formal systems is better than rough, you will recognize the following three ¶s as extremely crude and simplistic and are invited to treat them as **IYI** and skip or skim. A formal system of proof requires *axioms* and *rules of inference.* Axioms are basic propositions so obvious they can be asserted without proof. E.g. recall Euclid's Axioms or Peano's Postulates from school. Rules of inference, which are sometimes called the Laws of Thought, are the logical principles that justify deriving truths from other truths.[12] Some of the rules of inference are as simple as the Law of Identity, which basically holds that if anything is P, then it is P. Some others are more involved. For our purposes, two rules of inference are especially important. The first is known as the *Law of the Excluded Middle* (LEM). By LEM, a mathematical proposition P must be either true or, if not true, false.[13] The other big rule of inference involves the logical relation of *entailment,* meaning 'If . . . then' and often represented by the symbol '→'. The most obvious

[12] N.B. Significance notwithstanding, a proven truth in a formal system is technically known as a *theorem*—hence the Pythagorean Theorem, etc.

[13] **IYI** The 'or, if not true,' part is required in formal logic because of certain properties of the disjunctive operator 'or'. We'll do as little of this sort of arcana as possible. (Except while we're at it, let's confess that we're using 'LEM' in an informal way that also comprises the *principle of bivalence.* Letting 'LEM' connote the whole schmeer of Two-Valued Logic is fine for our purposes, but be advised that it's not 100% rigorous.)

rule of entailment is that (1) 'P → Q' and (2) 'P is true' license the conclusion (3) 'Q is true'. The one we're going to use a lot is the obverse of this rule and is usually called *modus tollens*; it holds that (1) 'P → Q' and (2) 'Q is false' license (3) 'P is false'. [14]

One reason why LEM and modus tollens are important to math is that they enable the method of Indirect Proof, also known as proof by *reductio ad absurdum* or sometimes just *reductio*. Here's how it works. Say you want to prove P. What you do is assume not-P and then show that not-P logically entails a contradiction like, say, 'Q & not-Q'. (By LEM, nothing can be both true and false, so the conjunction 'Q & not-Q' will always be false.) By modus tollens, if (1) not-P → (Q & not-Q) and (2) (Q & not-Q) is false, then (3) not-P is false; and, by LEM,[15] if not-P is false, then P must be true.

[14] Modus tollens (= Latin for 'method of denying') might not look like a universal rule unless you keep in mind that entailment, as a logical relation, has to do not with cause but with surety. *Necessary-* and *sufficient condition* are the terms in applied logic. If, for example, P is taken to mean 'is 5 feet tall' and Q to mean 'is at least 4 feet 11 inches tall,' then the purely logical meaning of 'P → Q' becomes evident: it really means 'If P is true then there is no way Q can be false'. Modus-tollenizing this into 'not-Q → not-P' is simply saying that if somebody is not 4′11″ then there is no way she can be 5′0″.

By the way, one other logical relation is going to be important far below in §5e and might as well get locked down here. It's the relation of *conjunction*, meaning 'and' and usually symbolized by '&' or '∧'. The big rule is that 'P & Q' is true only when P and Q are individually both true; if either one is false, the whole conjunction is false.

[15] **IYI** Well, this is technically not so much by LEM as by the definition of the truth-functional connective *not-*, which definition however either derives from LEM or (as some argue) is the same as LEM. N.B. also that some formal systems include the whole reductio transaction as an axiom, sometimes called the Law of Absurdity.

Many of the really great, famous proofs in the history of math have been reductio proofs. Here's an example. It is Euclid's proof of Proposition 20 in Book IX of the *Elements.* Prop. 20 concerns the primes, which—as you probably remember from school—are those integers that can't be divided into smaller integers w/o remainder. Prop. 20 basically states that there is no largest prime number. (What this means of course is that the number of prime numbers is really infinite, but Euclid dances all around this; he sure never says 'infinite'.) Here is the proof. Assume that there is in fact a largest prime number. Call this number P_n. This means that the sequence of primes (2, 3, 5, 7, 11, . . ., P_n) is exhaustive and finite: (2, 3, 5, 7, 11, . . ., P_n) is all the primes there are.[16] Now think of the number *R*, which we're defining as the number you get when you multiply all the primes up to P_n together and then add 1. *R* is obviously bigger than P_n. But is *R* prime? If it is, we have an immediate contradiction, because we already assumed that P_n was the largest possible prime. But if *R* isn't prime, what can it be divided by? It obviously can't be divided by any of the primes in the sequence (2, 3, 5, . . ., P_n), because dividing *R* by any of these will leave the remainder 1. But this sequence is all the primes there are, and the primes are ultimately the only numbers that a non-prime can be divided by. So if *R* isn't prime, and if none of the primes (2, 3, 5, . . ., P_n) can divide it, there must be some other prime that divides *R*. But this contradicts the assumption that

[16] You'll recollect from high school that ellipses inside a sequence or series mean 'all the relevant terms in between'; if they're at the end they mean 'and so on and so on w/o end'. This is an abbreviation that gets used in pure math a lot.

$(2, 3, 5, \ldots, P_n)$ is exhaustive of all the prime numbers. Either way, we have a clear contradiction. And since the assumption that there's a largest prime entails a contradiction, modus tollens dictates that the assumption is necessarily false, which by LEM[17] means that the denial of the assumption is necessarily true, meaning there is no largest prime. Q.E.D.

Please observe that primeness has nothing to do with the world; it concerns only relations between numbers. The Greeks were the real inventors of what we call math, because—again—they were the first people to treat numbers and their relations as abstractions rather than as properties of collections of real things. It's important to see what a leap this was. Just from the as it were fossil record, it's easy to see that math had its matrix in the concrete. As in the immediate concrete. Consider the facts that numbers are called 'digits' and that most counting systems—not just our base-10 but also the base-5 and -20 systems of prehistoric Europe—are clearly designed around fingers and toes. Or that we still talk about the 'leg' of a triangle or 'face' of a polyhedron, or that 'calculus' comes from the Greek word for pebble, etc. It's common knowledge that there were pre-Greek civilizations, as in e.g. the Babylonians and Egyptians, with a fair degree of sophistication in math; but theirs was an intensely practical math, used for surveying, trade and finance, navigation, & c. The Babylonians and Egyptians were, in other words, interested in the five oranges rather than the 5. It was the Greeks who turned math into an abstract system, a special symbolic language that allows people not just to describe the concrete

[17] Again, strictly speaking it's more complicated than that, but for our purposes LEM will do.

world but to account for its deepest patterns and laws. We owe them everything.[18] More to the point, the accomplishments of K. Weierstrass, G. Cantor, and R. Dedekind in modern set- and number theory are impossible to appreciate without understanding the hyperdimensional jump from math as a practical abstraction of real-world properties to math as a Saussurian "system of symbols . . . independent of the objects designated." Nor, though, is true appreciation possible without also considering the consequent "displacements that are incalculable . . . "; because the abstract math that's banished superstition and ignorance and unreason and birthed the modern world is also the abstract math that is shot through with unreason and paradox and conundrum and has, as it were, been trying to tie its shoes on the run ever since the beginning of its status as a real language. Re which, again, please keep in mind that a language is both a map of the world and its own world, with its own shadowlands and crevasses—places where statements that seem to obey all the language's rules are nevertheless impossible to deal with.

We can assume that most of the natural-language terrain is already familiar—but just as a mnemonic, consider the distance/levels involved between using 'tree' and 'rock' to designate actual trees and rocks and W. J. Clinton's excruciating semantics of 'inhale' or 'have sex'. Or parse the well-known 'I Am Lying' paradox (also a Greek invention). Or meditate on sentences like '"'Makes no sense' makes no sense" makes no sense' or '"Is, if it immediately follows its own quotation, false" is, if it immediately follows its own quotation, false'. You'll

[18] **IYI** including our abstraction-schizophrenia and slavery to technology and Scientific Reason, ultimately.

notice that these last three, like most paradoxical crevasses, involve either self-reference or regressus, two demons that have afflicted language since as far back as we might want to go.

Math is not exempt. And of course since mathematics is a totally abstract language, one whose lack of specific real-world referents is supposed to yield maximal hygiene, its paradoxes and conundra are much more of a problem. Meaning math has to really deal with them instead of just putting them in the back of its mind once the alarm goes off. Some dilemmas can be handled legalistically, so to speak, by definition and stipulation.[19] Easy example from high-school algebra: From the inarguable fact that the divisors in an equation of two fractions are equal if the numerators are—that is, if $\frac{x}{y} = \frac{x}{z}$ then $y = z$—it would seem that if $\frac{(x-5)}{(x-4)} = \frac{(x-5)}{(x-3)}$, then $(x-4) = (x-3)$, meaning $4 = 3$, which is clearly a crevasse. This is handled by decreeing that the only possible solution to $\frac{(x-5)}{(x-4)} = \frac{(x-5)}{(x-3)}$ is $x = 5$ (since 0 divided by anything yields the same 0, which obviously does not entail that $4 = 3$) and by stipulating that the theorem $(\frac{x}{y} = \frac{x}{z}) \rightarrow (y = z)$ holds only if $x \neq 0$.

Or here's a tricky one. We all remember repeating decimals, like the way $\frac{2}{3}$ is also .666 . . .. It turns out you can show that the repeating decimal .999 . . . is equal to 1.0 with only a couple

[19] One reason math texts are so abstruse and technical is because of all the specifications and conditions that have to be put on theorems to keep them out of crevasses. In this sense they're like legal documents, and often about as much fun to read.

wholly legal moves. To wit: If $x = .999\ldots$, then $10x = 9.999\ldots$; so then subtract x from $10x$:

$$\begin{array}{r} 9.9999999\ldots \\ -0.9999999\ldots \end{array}$$

and you get $9x = 9.0$ and thus $x = 1$. Is this specious or not? It depends how we treat the infinite sequence '.999 . . .,' like whether we choose to posit the existence of some number that's larger than .999 . . . but smaller than 1.0. Such a number would involve an *infinitesimal*, meaning a literally infinitely small mathematical entity. You might recall infinitesimals from college math. You may well however not recall—probably because you were not told—that infinitesimals made the foundations of the calculus extremely shaky and controversial for 200 years, and for much the same reason that Cantor's transfinite math was met with such howling skepticism in the late 1800s: nothing has caused math more problems—historically, methodologically, metaphysically—than infinite quantities. In many ways, the history of these ∞-related problems is the Story of Mathematics itself.

§1d. This warmup is all very fast and loose, of course. There are now some distinctions to nail down as we start approaching ∞ as an historical subject. The first is the obvious one between the infinitely large (= *transfinite*) and the infinitely tiny (= *infinitesimal*, $= \frac{1}{\infty}$). The second big difference is between ∞ as a feature of the physical world—as in questions like is the universe infinite, is matter infinitely divisible, does time have a beginning or end—and ∞ as an abstract mathematical entity or concept along the lines of function, number, primeness, and so on. There's already been a certain amount of warmup expended on the ontology of abstractions and whether/how math objects really exist, about which issues there is obviously still a whole library's worth of noodling that could be done. The important thing to keep straight is that the problems and controversies about ∞ that are going to concern us here involve whether infinite quantities can actually exist as mathematical entities.

The third distinction may at first seem picayune. It concerns ∞-related words like 'quantity' and 'number'. These have a weird and confusing double connotation, the same way words like 'length' or 'ounce' do. A section of rope has a certain length but is also sometimes called 'a length of rope'; a certain quantity of drugs that weighs one gram is also called 'a gram of drugs'. In just the same way, 'quantity' and 'number' can function both predicatively—that is, as satisfying the questions 'how much?' or 'how many?' of a certain thing—and as regular nouns denoting the thing described. It can thus be ambiguous, when a term like 'infinite number' is used, whether it is being used predicatively ('There are an infinite number of primes') or nominatively ('Cantor's first infinite number is $\aleph_0$'). And the difference is important, because the predicative usage of '∞' can be fuzzy and mean just 'indefinitely large' or 'really, really big,' whereas after Dedekind and Cantor the nominative has a very specific, if abstract, denotation.

In certain respects, the power and maybe even whole raison d'être of the language of math is that it's designed to be so clean and nonconnotative that it avoids ambiguities like those just above. Trying to express numerical quantities and relations in natural language—to translate mathematical propositions into English and vice versa—often causes trouble.[20] A favorite of Dr. Goris's was the old saw about three men who check into a motel late at night. There's only one available room left, and it costs $30, and the men decide to each chip in $10 and share it, but when they get up to the room it's a disgusting mess—apparently there was some mixup and the room never got cleaned after the last people checked out—and understandably the men call down to the manager to complain. Certain narrative details and flourishes can be omitted. The thrust is that it's late, and Housekeeping is long gone for the day, and there's no other room to move them to, so after a certain amount of back and forth the manager agrees to knock $5 off the price of the room and to supply clean linens, and he sends a bellboy up to the room with the linens and towels and the $5 refund in the form of five $1 bills. Etc. etc., with the point being that there's five $1 bills and three guys, so what the guys (who've mysteriously mellowed) do is they each take back one dollar and let the bellboy keep the remaining $2 as a tip. So each man originally paid $10 and got $1 back, meaning each paid $9, which adds up to $27, and the bellboy has the other $2, which all together sums to $29, so where's the other dollar? In which problem the point is that the verbiage (of which there was considerably more in the Dr. G. version—he had a whole running year-long epic about these three men and their various bios and travails and the different mathematical conundra they were always bumbling into) lulls you into fuzzily trying to calculate $(30 - 3) + 2$ instead of the apposite $(30 - 5) + (3 + 2)$,

[20] You will remember how especially unpleasant Word Problems were, in math classes.

resulting in much confusion and mirth and possible Extra Credit.

There are all sorts of interlinguistic teasers like this. The ones that aren't solvable become actual paradoxes, some of which are profound. It should be unsurprising, since ∞ is both the ultimate abstraction and inveterately fuzzy, that it figures in many such paradoxes. Like take the ideas that there is no last or largest integer and that time extends infinitely forward. Then imagine a perfectly constructed and durable desk lamp with a big red On/Off button, and imagine that this morning the lamp is off but at 4:30 P.M. CST it will be turned on, then at 4:30 P.M. tomorrow it will be turned back off, but then back on again at 4:30 the following day, and so on, every day, for the rest of time; and now contemplate whether, after an infinite number of days, the lamp will be on or off. You might recall from college math[21] that this is actually a Word Problem involving something called a *divergent infinite series*, more specifically the Grandi Series, $1 - 1 + 1 - 1 + 1 - 1 + 1 \ldots$, which series sums to 0 if we compute it as $(1 - 1) + (1 - 1) + (1 - 1) + \ldots$ but sums to 1 if we compute it as $1 + (-1 + 1) + (-1 + 1) + (-1 + 1) \ldots$, with

[21] **IYI** Let's explicitize at the outset that the 'you might recall's and 'it goes without saying's and so on are not tics but rhetorical gambits whose aim is to reduce annoyance in those readers who are already familiar with whatever's being discussed. No particular experience or recall of college math is actually required for this booklet; but it seems only reasonable to assume that some readers will have strong math backgrounds, and only polite to acknowledge this from time to time. As was briefly mentioned in the Foreword, the rhetoric of tech writing is fraught with conundra about various different readers' expertise-levels and confusion-v.-annoyance curves. None of this is your problem, of course—at least not directly.

the kicker being that, since both computations are mathematically licit, the series' 'real' sum appears to be both 1 and not-1, which by LEM is impossible. You may or may not further recall, however,[22] that the Grandi Series happens to be a particular subtype of divergent infinite series known as an *oscillating series*, and that as such it's an object lesson in stipulation for partial sums (symbolized s_n), with the relevant symbolism being '$1 + \Sigma(-1)^n$ where $s_n = 0$ for even n and $s_n = -1$ for odd n'—which symbolism appears so legalistically arcane and clunky precisely because it has to be designed to avoid crevasses like the Lamp Paradox.

Or there are antinomies that revolve around ∞ not as a natural-language concept or a numerical *terrain vague* but merely as a feature of geometry, and can be represented in simple pictures, and can't just be stipulated away. Take the matter of points and lines. It's a given that any line includes infinitely many points. A point, recall from school, is "an element of geometry having position but no extension," meaning that a point is an abstraction, pure location. But if a line is composed entirely of points, and points have no extension, how can a line have extension? Which all lines by definition do. The answer seems to have to do with ∞, but how can even $\infty \times 0$ equal anything more than 0?

[22] The following clause isn't quite **IYI**, but to follow it entirely you probably would need to have had some college math, which if you haven't, don't worry about anything more here than the way the symbolism has to get freighted with stipulations. (**IYI** And if you have had Calc I-II and are noticing that our symbolism is somewhat nonstandard, again don't fret—it's mostly because none of the relevant symbols has been defined yet.)

Here's an even worse one. All you need is Euclid and a ruler. Draw a line like so:

P Q R

where the line segment PQ is three times longer than the segment QR. Since line segments are composed of points, it stands to reason that there should be three times as many points on PQ as on QR. But it turns out there are exactly as many in both. You can see it. Turn the line into the right triangle QPR by rotating PQ up and over so that P is right over R and then drawing the line segment PR:

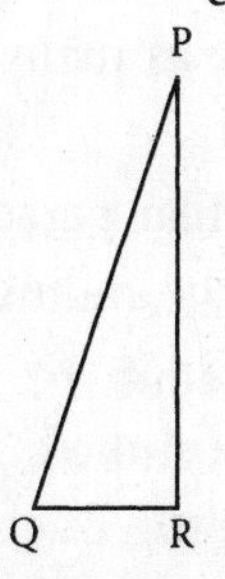

Then recall that, via Euclid's Parallel Axiom,[23] through any point on the segment PQ there will exist exactly one line parallel to the segment PR:

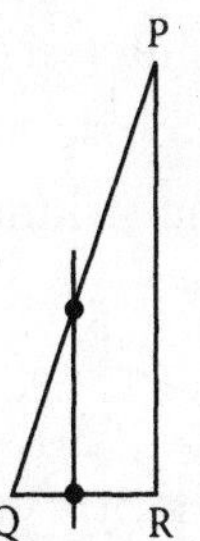

and that this line will hit the segment QR at exactly one point. The same is true for every single point on PQ—simply

[23] **IYI** = Def. 23 in Book I of the *Elements*.

draw a line parallel to PR that hits PQ at this point, and the line will hit QR at one and only point:

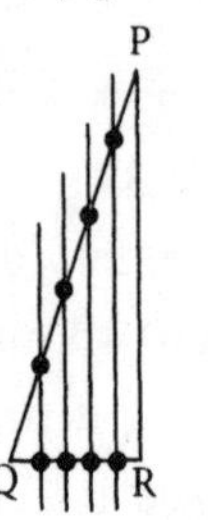

with no duplications and no points left over, meaning that for every point on PQ there is a corresponding point on QR, meaning there are exactly as many points on QR as on PQ, even though PQ = 3(QR).

You can generate a similar paradox with circles and make a concentric duo where the radius of the big circle is twice that of the one inside.[24] Since any circle's circumference is a direct function of its radius, the big one's circumference will be twice the size of the little one's. And a circumference is also a line, so there ought to be twice as many points on the big circle's circumference. But no: since the two circles have a common center, simply drawing in some sample radii establishes that any radius that intersects the larger circle at a point N will intersect the little circle at one and only one corresponding point N_1, with no redundancies or remainder:

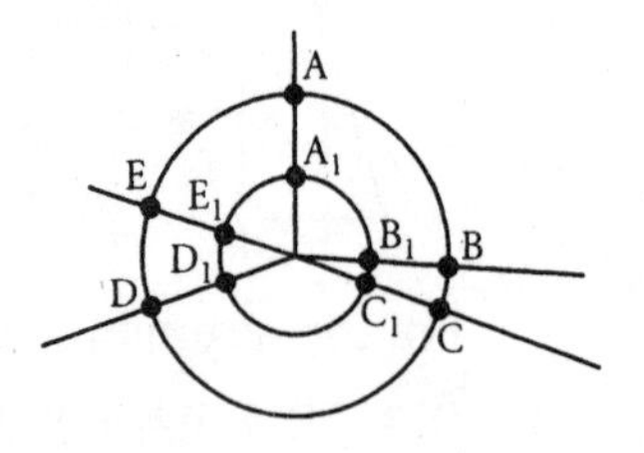

[24] **IYI** This one is related to something called Aristotle's Wheel, which is a whole story in itself.

thereby demonstrating that the number of points on both circles' circumferences is the same.

These are real problems, not just irksome or counterintuitive but mathematically profound. G. F. L. P. Cantor solved them all, more or less. But of course the natural-language 'solve' can mean different things. As mentioned, one way in math to take care of destabilizing problems is to legislate them out of existence—by banishing certain kinds of math entities and/or by loading theorems with stipulations and exclusions designed to head off crazy results. Prior to the invention of transfinite math, this was the way most paradoxes of ∞ were handled. You 'solved' them by first fudging the distinction between a paradox and a contradiction, and then by applying a kind of metaphysical reductio: if allowing infinite quantities like the number of points on a line or the set of all integers led to paradoxical conclusions, there must be something inherently wrong or nonsensical about infinite quantities, and thus ∞-related entities couldn't really 'exist' in a mathematical sense. This was essentially the argument deployed against, e.g., the famous Paradox of Galileo in the 1600s. Here's the Paradox of Galileo. Euclid's 5th Axiom dictates that "the whole is always greater than the part," which seems pretty unassailable. It's also obvious that while every perfect square (viz. 1, 4, 9, 16, 25, . . .) is an integer, not every integer is a perfect square. In other words the set of all perfect squares is but a part of, and so by Euclid's 5th smaller than, the set of all integers. The trouble is that the same sort of equality-via-correspondence we saw with PQ/QR and the two circles can be set up here. Because while not every integer is a perfect square, every integer *does* happen to be the square root of a perfect square—2 of 4, 3 of 9, 4 of 16, 912 of 831,744, and so on. Pictorially, you can line up the two sets and demonstrate a perfect and inexhaustible one-to-one correspondence between their members[25]:

$$
\begin{array}{ccccccccccc}
1 & 2 & 3 & 4 & 5 & \cdots & 911 & 912 & \cdots & n & \cdots \\
\updownarrow & \updownarrow & \updownarrow & \updownarrow & \updownarrow & & \updownarrow & \updownarrow & & \updownarrow & \\
1 & 4 & 9 & 16 & 25 & \cdots & 829{,}921 & 831{,}744 & \cdots & n^2 & \cdots
\end{array}
$$

The upshot of the Paradox of Galileo is thus that Euclid's 5th Axiom—an indispensable part of basic math, not to mention an obvious truth borne out by every single kind of set we can ever see or count—is contradicted by the infinite sets of all integers and all perfect squares. Given this situation, there are two ways to go. The standard way, as mentioned, is to declare infinite sets the math equivalent of unicorns or the 'nothing' Alice sees on the road.[26] The other—which is revolutionary, both intellectually and psychologically—is to treat Galileo's paradoxical equivalence not as a contradiction but as a description of a certain new kind of mathematical entity that's so abstract and strange it doesn't conform to math's normal rules and requires special treatment. I.e., it is to point out (as did guess who) that "The fundamental flaw of all so-called proofs of the impossibility of infinite numbers is that they attribute to these numbers all the properties of finite numbers, whereas the infinite numbers . . . constitute an entirely new type of number, a type whose nature should be an object of research instead of arbitrary prejudice."

[25] **IYI** As we'll see in §7, a one-to-one correspondence between their members is now actually the definition of equality between two sets.

[26] **IYI** That's a little crude. The real way math neutered ∞ was based on a certain metaphysical distinction drawn by Aristotle in opposition to Zeno, all of which we'll get to.

Except on the other hand such an attitude could be not revolutionary but merely insane.[27] Rather like taking the fact that nobody's ever once seen a unicorn and claiming that this is not an indication that unicorns don't really exist but rather evidence that unicorns constitute a whole new type of animal with the unique property of invisibility. Here of course we get the Fine Line Between Brilliance and Madness that modern writers/filmmakers dine out on. The truth is that all manner of strange, non-directly-observable entities such as 0, negative integers, irrational numbers, etc. originally entered math under the same sort of insanity/incoherence cloud but are now totally accepted, even essential. At the same time, there have been plenty of other innovations that really were insane or unworkable and got laughed out of town, mathematically speaking, and we laymen never hear of them.

Just because something's a Fine Line doesn't mean it isn't a line, though. Mathematical thinking is abstract, but it's also thoroughly private-sector and results-oriented. The difference between a brilliant, revolutionary mathematical theory and a wacko one lies, therefore, in what-all can be done with it, in whether or not it yields significant results. Here's G. H. Hardy explaining 'significant results':

> We may say, roughly, that a mathematical idea is 'significant' if it can be connected, in a natural and illuminating

[27] It's only fair to point out that Cantor made his share of wacky-sounding pronouncements in the controversy over infinite sets, many with an ominously grandiose religious flavor, as in "I entertain no doubts as to the truth of the transfinites, which I have recognized with God's help," or "The fear of infinity is a form of myopia that destroys the possibility of seeing the actual infinite, even though it in its highest form has created and sustains us."

way, with a large complex of other mathematical ideas. Thus a serious mathematical theorem, a theorem which connects significant ideas, is likely to lead to important advances in mathematics itself and even in other sciences.

G. F. L. P. Cantor's theories of infinite sets and transfinite numbers end up being significant in precisely this way. Part of the reason is that Cantor was an extremely fine working mathematician and derived ingenious proofs of the important formal features that made his ideas real theories instead of just bold hypotheses. But there are other reasons, too. Galileo himself had hypothesized that the real upshot of his Paradox was that "the attributes 'equal,' 'greater,' and 'less' are not applicable to infinite, but only to finite quantities." No one took this seriously, though, and not because of stupidity—math does not tend to be rife with stupid or closed-minded people. The time was literally not right for Galileo's suggestion; nor were there yet the right mathematical tools for making it a real theory even if Galileo'd wanted to . . . which he didn't, from which fact it would be wrong to conclude that he wasn't just as farsighted or brilliant as G. Cantor. Like most of the giants who revolutionize math or science, Cantor was 100% a man of his time and place, and his accomplishments were the usual conjunction of extraordinary personal brilliance and courage[28] and just the right context of general

[28] Naturally, the significant results that legitimize a mathematical theory take time to derive, and then even more time to be fully accepted, and of course throughout this time the Insanity-v.-Genius question remains undecided, probably even for the mathematician himself, so that he's developing his theory and cooking his proofs under conditions of enormous personal stress and doubt, and sometimes isn't even vindicated in his own lifetime, etc.

problems and conditions that, in hindsight, tend to make intellectual advances seem inevitable and their authors almost incidental.

To put it another way, math is pyramidical; Cantor didn't just suddenly come out of nowhere. Real appreciation therefore requires understanding the concepts and problems that gave rise to set theory and made transfinite math significant in Hardy's sense. This takes a while, but since the discussion is itself pyramidical we can proceed in a more or less ordered way, and the whole thing won't be nearly as abstract and discursive as this warmup.

§2a. Now we're really starting. There are two ways to trace out the context of Cantorian set theory. The first is to talk about the abstract intertwined dance of *infinity* and *limit* throughout math's evolution. The second is to examine math's historical struggle with representing *continuity*, meaning the smooth-flowing and/or densely successive aspects of motion and real-world processes. Anyone with even the vaguest memory of college math will recall that continuity and ∞/limit are pretty much the fundament of the calculus, and might also recall that they have their general origins in the metaphysics of the ancient Greeks and their particular matrix in Zeno of Elea (c. 490–435 BCE, who died with his teeth literally still in the ear of Elea's despotic ruler Nearchus I (long story)), whose eponymous paradoxes pulled the starter-rope on everything.

A few Attic facts at the outset. First, Greek math was abstract all right, but it had its roots deep in Babylo-Egyptian praxis. There is no real difference, for the Greeks, between arithmetical entities and geometric figures, between e.g. the number 5 and a line five units long. Nor, second, are there any clear distinctions for the Greeks between mathematics, metaphysics, and religion; in many respects they were all the same thing. Third, our own age and culture's dislike of limits—as in 'a limited man,' 'IF YOUR VOCABULARY IS LIMITED, YOUR CHANCES FOR SUCCESS ARE LIMITED,' etc.—would have been incomprehensible to the ancient Greeks. Suffice to say they liked limits a *lot*, and a straightforward consequence of this is their distaste for/distrust of ∞. The Hellenic term *to apeiron* means not only infinitely long/large but also undefinable, hopelessly complex, the that-which-cannot-be-handled.[1]

To apeiron also and most famously refers to the unbounded natureless chaos from which creation sprang. Anaximander (610–545 BCE), the first of the pre-Socratics to use the term in his metaphysics, basically defines it as "the unlimited substratum from which the world derived." And the 'unlimited' here means not only endless and inexhaustible but formless, lacking all boundaries and distinctions and specific qualities. Sort of the Void, except what it's primarily devoid of is form.[2] And this, for the Greeks, is not good. Here's a definitive quotation from Aristotle, that font of definitive quotations: "[T]he essence of the infinite is privation, not perfection but the absence of limit." The point being that in abstracting away all limits to get ∞ you are throwing the baby out as

[1] **IYI** The term *to apeiron* apparently originated in Greek tragedy, where it referred to garments or binds 'in which one is entangled past escape'.

[2] **IYI** Probably worth observing here that *Genesis* 1:2's "The earth was without form and void" is a thoroughly Greek way to characterize pre-Creation.

well: no limit means no form, and no form means chaos, ugliness, a mess. Note thus Attic Fact Four, the ubiquitous and essential aestheticism of the Greek intellect. Messiness and ugliness were the ultimate *malum in se*, the sure sign that something was wrong with a concept, in much the same way that disproportion or messiness was impermissible in Greek art.[3]

Pythagoras of Samos (570–500 BCE) is crucial in all kinds of ways to the history of ∞. (Actually it's more accurate to say 'the Divine Brotherhood of Pythagoras' or at least 'the Pythagoreans,' because ∞-wise the man was less important than the sect.) It was Pythagorean metaphysics that explicitly combined Anaximander's *to apeiron* with the principle of limit (= Gr. *peras*) that lends structure and order—the possibility of form—to the primal Void. The Divine Brotherhood of P., who as is well-known made a whole religion of Number, posited this limit as mathematical, geometric. It is the operations of *peras* on *to apeiron* that produce the geometrical dimensions of the concrete world: *to apeiron* limited once produces the geometric point, limited twice produces the line, three times the plane, and so on. However odd or primitive this might seem, it was extremely important, and so were the Pythagoreans. Their *peras*-based cosmology meant that the genesis of numbers was the genesis of the world. The D.B.P. were, yes, legendarily eccentric, as in their seasonal rules about sex or Pythagoras's pathological hatred of

[3] N.B. that this Hellenic aestheticism has never died out in math, as in the way a great proof or method is called 'elegant,' or Hardy's *Apology*'s oft-quoted "Beauty is the first test; there is no permanent place in the world for ugly mathematics."

legumes. But they were the first people to regard, and revere, numbers as abstractions. The centrality of the number 10 to their religion, for example, was based not on finger- or toe-factors but on 10's status as the perfect sum of $1 + 2 + 3 + 4$.

The D.B.P. were also the first philosophers explicitly to address the metaphysical relation between abstract mathematical realities and concrete empirical realities. Their basic position was that mathematical reality and the concrete world were the same, or rather that empirical reality was a sort of shadow or projection of abstract math.[4] Moreover, many of their arguments for the primacy of number were based on the observed fact that purely formal mathematical relationships had striking implications for real-world phenomena, a famous example being how the D.B.P. abstracted the Golden Mean ($\frac{x}{1+x} = \frac{1}{x}$, which solves to roughly $\frac{55}{34}$) from seashells' whelks and trees' rings and promulgated its use in architecture. As mentioned, some of these math/world connections had been known to earlier cultures like the Egyptians—or maybe rather 'used by them' would be better, since the Egyptians had zero interest in what the connections actually were, or meant. A couple more examples. In practice, the Egyptians had used what we now call the Pythagorean

[4] **IYI** In case you're noticing that this is a pretty Platonic-sounding description, be apprised that even though Plato lived a century after Pythagoras, there's good evidence that he came into close contact with later members of the D.B.P. during his travels through Greek-held southern Italy, and that their metaphysics of math underlie Plato's own Theory of Forms, re which see just below.

Theorem in engineering and surveying along the Nile; but it was Pythagoras who made it an actual Theorem, and proved it. Plenty of pre-Greek cultures also played music, but it was the D.B.P. who discovered the concepts of the octave, the perfect fifth, etc., by observing that certain musical intervals always corresponded to certain ratios in the lengths of plucked strings—2 to 1, 3 to 2, and so on. Since strings were lines and lines were geometric/mathematical entities,[5] the ratios of strings' lengths was the same as the ratios of integers, a.k.a. rational numbers, which happen to be the fundamental entities of Pythagorean metaphysics.

Etc. etc., the point being that the D.B.P.'s attempts to articulate the connections between mathematical reality and the physical world were part of the larger project of pre-Socratic philosophy, which was basically to give a rational, nonmythopoeic account of what was real and where it came from. Maybe even more important than the D.B.P., ∞-wise, is the protomystic Parmenides of Elea (c. 515–? BCE), not only because his distinction between the 'Way of Truth' and 'Way of Seeming' framed the terms of Greek metaphysics and (again) influenced Plato, but because Parmenides' #1 student and defender was the aforementioned Zeno, the most fiendishly clever and upsetting Greek philosopher ever (who can be seen actually kicking Socrates' ass, argumentatively speaking, in Plato's *Parmenides*). Zeno's arguments for Parmenidean metaphysics took the form—again as mentioned—of some of the most profound and nutcrunching paradoxes in world history. In support of these crunchers' relevance

[5] The Egyptians' own concept of 'line,' on the other hand, had been just a stretched rope at the edge of somebody's property.

to our overall purpose, here is another nice B. Russell quotation:

> In this capricious world, nothing is more capricious than posthumous fame. One of the most notable examples of posterity's lack of judgment is the Eleatic Zeno [. . .], who may be regarded as the founder of the philosophy of infinity. He invented four arguments, all immeasurably subtle and profound, to prove that motion is impossible, that Achilles can never overtake the tortoise, and that an arrow in flight is really at rest. After being refuted by Aristotle, and by every subsequent philosopher from that day to our own, these arguments were reinstated, and made the basis of a mathematical renaissance, by a German professor, who probably never dreamed of any connection between himself and Zeno.

For the record, Parmenides' metaphysics—which is even wilder than the D.B.P.'s, and in retrospect seems more like Eastern religion than Western philosophy—is describable as a kind of static monism,[6] and Zeno's Paradoxes (of which there are really more than four) are accordingly directed against the reality of (1) plurality and (2) continuity. For present purposes we are concerned with (2), which for Zeno takes the form, as Russell mentions, of regular physical motion.

Zeno's basic argument against the reality of motion is known as the Dichotomy. It looks very simple and is deployed in two of his most famous paradoxes, "The Racetrack" and "Achilles v. the Tortoise." The Dichotomy later gets used and discussed, with all sorts of different setups and

[6] = roughly 'All Is One' + 'Nothing Changes'.

agendas, by Plato, Aristotle, Agrippa, Plotinus, St. Thomas, Leibniz, J. S. Mill, F. H. Bradley, and W. James (to say nothing of D. Hofstadter in *Gödel, Escher, Bach*). It runs thus.[7] You're standing at a corner and the light changes and you try to cross the street. Note the operative 'try to'. Because before you can get all the way across the street, you obviously have to get halfway across. And before you can get halfway across, you have to get halfway to that halfway point. This is just common sense. And before you can get to the halfway-to-the-halfway-point point, you obviously have to get halfway to the halfway-to-the-halfway-point point, and so on. And on. Put a little more sexily, the paradox is that a pedestrian cannot move from point A to point B without traversing all successive subintervals of AB, each subinterval equaling $\frac{AB}{2^n}$ where n's values compose the sequence (1, 2, 3, 4, 5, 6, . . .), with the ' . . . ' of course meaning the sequence has no finite end. Goes on forever. This is the dreaded *regressus in infinitum*, a.k.a. the Vicious Infinite Regress or VIR. What makes it vicious here is that you're required to complete an infinite number of actions before attaining your goal, which—since the whole point of 'infinite' is that there's no end to the number of these actions—renders the goal logically impossible. Meaning you can't cross the street.

The standard way to schematize the Dichotomy is usually:

(1) In order to traverse the interval AB you must first traverse all the subintervals $\frac{AB}{2^n}$ where $n = 1, 2, 3, 4, 5, 6, \ldots$.

[7] The following should not be classed **IYI** even if you know the Dichotomy already, since the discussion here is rather specially tailored.

(2) There are infinitely many such subintervals.
(3) It is impossible to traverse infinitely many subintervals in a finite amount of time.
(4) Therefore, it is impossible to traverse AB.

It goes without saying that the interval AB doesn't have to be a very wide street, or even a street at all. The Dichotomy applies to any kind of continuous motion. Dr. G. in class used to like to run the argument in terms of the DUI-like movement of your finger from your lap to the tip of your nose. And of course, as anybody who's ever successfully crossed a street or touched his nose is aware, there has got to be something fishy about Zeno's argument. Finding and articulating that fishiness is a whole other matter. We have to be careful, too; there's more than one way to be wrong. If you've had some college math, for instance, it may be tempting to say that the Dichotomy's step (2) conceals a simple fallacy, namely the assumption that the sum of an infinite series must itself be infinite. You might recall that step (1)'s $\frac{AB}{2^n}$ is simply another way to represent the geometric series $\frac{1}{2^1} + \frac{1}{2^2} + \frac{1}{2^3} + \frac{1}{2^4} + \cdots$, and that the correct formula for finding the sum of this geometric series is $\frac{a}{(1-r)}$, where a is the series' first term and r is the common ratio, and that here a is $\frac{1}{2}$ and so is r, and $\frac{\frac{1}{2}}{(1-\frac{1}{2})} = 1$, in which case it appears that streets can be crossed and noses touched with no problem, and thus that the Dichotomy is really just a tricky Word Problem and not a paradox at all, except maybe for civilizations too crude and benighted to know the formula for summing a geometric series.

Except this response won't do. Leave aside for the moment whether it's technically correct. What matters is that it's trivial; it represents what philosophers would call an *impoverished view* of Zeno's problem. For whence exactly $\frac{a}{(1-r)}$ as a formula for summing this geometric series? I.e., is the formula just a bit of lawyerly semantics designed to define certain paradoxes out of existence, or is it mathematically significant in Hardy's sense of 'significant'? And how do we determine which it is?

Weirdly, the more standard classroom math you've had, the harder it's going to be to avoid answering in an impoverished way. Such as, e.g., validating $\frac{a}{(1-r)}$ by observing, in the best Calc II tradition,[8] that the relevant geometric series here is a particular subtype of *convergent infinite series*, and that the sum of such a series is defined as the limit of the sequence of its partial sums (that is, if the sequence $s_1, s_2, s_3, \ldots, s_n, \ldots$ of a series' partial sums tends to a limit S, then S is the sum of the series), and that sure enough, w/r/t the above series, $\lim_{n \to \infty} s_n = 1$, so $\frac{a}{(1-r)}$ works just fine . . . in which case you will once again have answered Zeno's Dichotomy in a way that is complex, formally sexy, technically correct, and deeply

[8] meaning, again, that the following will make 100% sense only if you've had the relevant math, which again if you haven't don't worry—what's important is the overall form of the reasoning, which you can get without knowing the specific terms/symbols. (**IYI** In fact, the terms/symbols in play here are all going to get defined below, but not until we really need them.)

trivial. Along the lines of 'Because it's illegal' as an answer to 'Why is it wrong to kill?'

The trouble with college math classes—which classes consist almost entirely in the rhythmic ingestion and regurgitation of abstract information, and are paced in such a way as to maximize this reciprocal data-flow—is that their sheer surface-level difficulty can fool us into thinking we really know something when all we really 'know' is abstract formulas and rules for their deployment. Rarely do math classes ever tell us whether a certain formula is truly significant, or why, or where it came from, or what was at stake.[9] There's clearly a difference between being able to use a formula correctly and really knowing how to solve a problem, knowing why a problem is an actual mathematical *problem* and not just an exercise. In this regard see yet another part of the B. Russell ¶ on Zeno,[10] this time with emphases supplied:

> Zeno was concerned, as a matter of fact, with three problems, each presented by motion, but each more abstract than motion, and capable of purely arithmetical treatment. These are the problems of the infinitesimal, the infinite, and continuity. *To state clearly the difficulties involved, was to accomplish perhaps the hardest part of the philosopher's task.*

[9] And, of course, rarely do students think to ask—the formulas alone take so much work to 'understand' (i.e., to be able to solve problems correctly with), we often aren't aware that we don't understand them at all. That we end up not even knowing that we don't know is the really insidious part of most math classes.

[10] **IYI** Russell gets quoted so much here because his prose on all this is extremely pellucid and fine—plus notice the way he, like the Greeks, makes no real distinction between math and philosophy.

And '$\frac{a}{(1-r)}$' fails, without a great deal of context and as it were motivation, to state clearly the difficulties involved. Stating these difficulties clearly is, in fact, the whole and only difficulty involved here (and if you can now feel the slight strain and/or headache starting, you'll know we're in Zeno's real territory).

First, to save at least 10^3 words, have a refresher-type look at the following two rough graphs, one of the divergent sequence[11] for 2^n and the other of the convergent sequence for $\frac{1}{2^n}$:

Exhibit 2a(1)

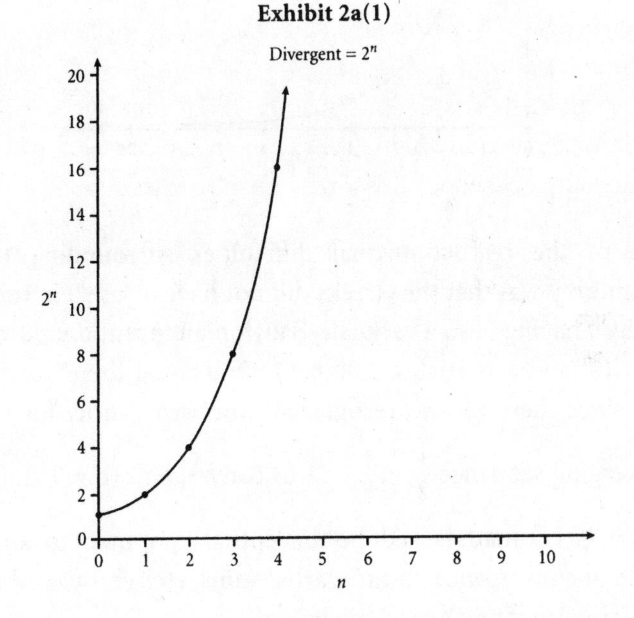

[11] **IYI** meaning one that doesn't have a finite limit. Convergence and divergence might not make complete sense until we talk about limits in §3c. All you need at this point is a rough idea of what divergence-v.-convergence involves, which the exhibits are supposed to enable.

Exhibit 2a(2)

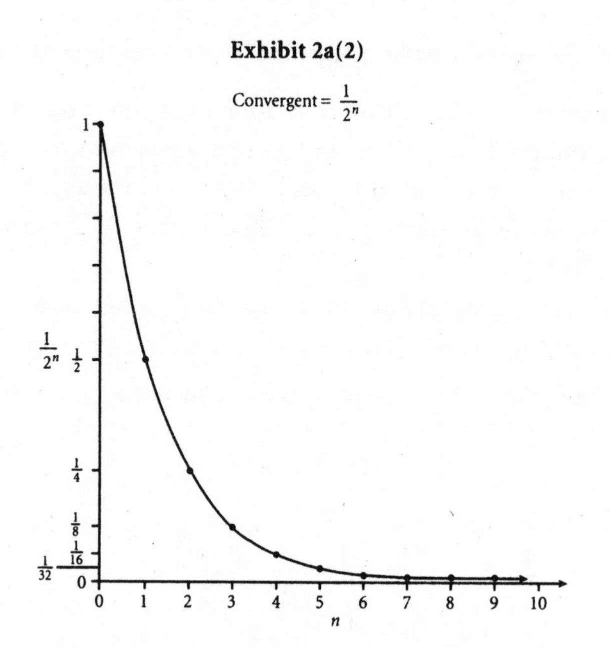

One of the real contextual difficulties surrounding the Dichotomy was that the Greeks did not have or use 0 in their math (0 having been a very-late-Babylonian invention, purely practical and actuarial, c. 300 BCE). One could therefore say that since there was no recognized number/quantity for the converging sequence $\frac{1}{2}, \frac{1}{4}, \frac{1}{8}, \ldots$ to converge to (q.v. Exhibit 2a(2)), Greek math lacked the conceptual equipment to comprehend convergence, limits, partial sums, etc. This would be true in a way,[12] and not wholly trivial.

[12] **IYI** The ways in which it's not 100% true involve Eudoxus of Cnidos, who tends to get even less press than Zeno—q.v. §2d below.

Less trivial still is the aforementioned Greek dread of *to apeiron.* Zeno was the first philosopher to use ∞'s black-hole-like logical qualities as an actual argumentative tool, viz. the Vicious Infinite Regress, which even today gets used in logical arguments as a reductio-grade method of proof. Example: In epistemology, the VIR is the easiest way to refute the common claim that in order to really know something you have to know that you know it. Like most VIR proofs, there's an evil-edged fun to this one. Let the variable x denote any fact or state of affairs precedable by the expletive 'that,' and restate the original claim as (1) 'In order to know that x, you must know that you know that x'. Since the whole expletive phrase 'that you know that x' qualifies as a fact or state of affairs, in the proof's next step you can simply expand the denotation of x so that now $x =$ [you know that x] and then substitute it into the original claim, resulting mutatis mutandis in (2) 'In order to know that [you know that x], you must know that you know that [you know that x],' the next wholly valid x-extension of which then yields (3) 'In order to know that [you know that you know that x], you must know that you know that [you know that you know that x],' and so on, ad inf., requiring you to satisfy an endless number of preconditions for knowing anything.

(**IYI** The VIR is so powerful a tool that you can easily use it to annoy professional competitors or infuriate your partner in domestic conflicts, or (worse) to drive yourself crazy in bed in the morning over, e.g., any kind of relation between two things or terms, like when we say that 2 and 4 are related by the function $y = x^2$, or that if clouds cause rain then clouds and rain stand in a causal relation. If you consider the idea in the abstract and ask, w/r/t any relation, whether this relation is itself related to the two terms it relates, the answer is inescapably yes (since it's impossible to see how a relation can connect two terms unless it has its own relation to each one, the way a bridge between two riverbanks has got to be connected to each bank), in which case the relation between, say, clouds and rain actually entails two more relations—viz. those between (1) clouds and the relation and (2) rain and the relation—each of which latter relations obviously then entails two more on either side, and so on, ad inf. . . . which is not a fun or productive abstract path to venture down in the A.M. at all, especially since the geometric series of relations here is divergent rather than convergent, and as such it's connected to all kinds of especially dreadful and modern divergent series like the exponential doublings of cancer, nuclear fission, epidemiology, & c. Worth noticing also is that hideous divergent VIRs like those above always involve the metaphysics of abstractions, such as 'relation' or 'knowledge'. It's like some fissure or crevasse always opens up in the move from particular cases of knowing/relating to knowledge/relation *in abstractus.*)

Zeno himself is almost fetishistically attached to the divergent VIR and uses it in several of his lesser-known paradoxes. Here is a specifically anti-Pythagorean Z.P.[13] contra the idea

[13] **IYI** All anybody knows of Zeno's Paradoxes is from secondary sources, since either Zeno didn't write anything or it's all been lost. The above paradox appears most famously in §209a of Aristotle's *Physics*—and notice how this one too revolves around issues in the ontology of abstractions, particularly in the move from steps (1) to (2).

Also: If you are the sort of person who can keep seemingly irrelevant things in your head for several pages, then observe now that the exact same metaphysical slipperinesses in this Z.P. will reappear w/r/t the Dichotomy in questions about what exactly a mathematical point is, like whether a point is a geometrical abstraction, or an actual physical location, or both.

that anything can really be in a particular location, which in simplified form is schematized:

(1) Whatever exists is in a location.
(2) Therefore, location exists.
(3) But by (1) and (2), location must be in a location, and
(4) By (1)–(3), location's location must itself be in a location, and . . .
(5) . . . So on ad inf.

This one's rather easier to see the gears of, since the true Russellian difficulty here is some slipperiness around 'to exist'. Actually, since ancient Greek didn't even have a special verb for existence, the relevant infinitive is the even slipperier 'to be'. On purely grammatical grounds, Zeno's argument can be accused of the classic Fallacy of Equivocation,[14] since 'to be' can have all sorts of different senses, as in 'I am frightened' v. 'He is a Democrat' v. 'It is raining' v. 'I AM THAT I AM'. But you can see that pressing this case will (once again) lead quickly to the paradox's deeper questions, which questions here are (again once again) metaphysical: what exactly do these different senses of 'to be' mean, and in particular what does the more specialized sense 'to exist' mean, i.e. what sorts of things really do exist, and in what ways, and are there different kinds of existence for different kinds of things, and if there are then are some kinds of existence more basic or substantial than other kinds? & c.

[14] As in:

(1) Curiosity killed the cat.
(2) The World's Largest Ball of Twine is a curiosity.
(3) Therefore the World's Largest Ball of Twine killed the cat.

You'll have noticed that we've run up against these sorts of questions a dozen times already and we're still 2,000+ years away from G. Cantor. They are the veritable bad penny in the Story of ∞, and there's no way around them if you don't want just a bunch of abstract math-class vomitus on transfinite set theory. Deal. Right now is the time for a sketch of Plato's One Over Many argument, which is the classic treatment of just these questions as they apply to the related issue of predication.

You might recall the O.O.M., too, from school, in which case relax because this won't take long. For Plato, if two individuals have some common attribute and so are describable[15] by the same predicate—'Tom is a man'; 'Dick is a man'—then there is something in virtue of which Tom and Dick (together with all other referents of the predicate nominative 'man') have this common attribute. This something is the ideal Form Man, which Form is what really, ultimately exists, whereas individual men are just temporal appearances of the Form, with a kind of borrowed or derivative existence, like shadows or projected images. That's a very simplified version of the O.O.M., but not a distorted one—and even at this level it should not be hard to see the influences of Pythagoras and Parmenides on Plato's ontological Theory of Forms, which the O.O.M. is an obvious part of.

Here's where the truth gets a little complicated. As seems to happen a lot, the complication involves Aristotle. It's true that the first mention of the O.O.M. is in Plato's *Parmenides*, but in fact what made the argument famous is Aristotle's

[15] via the special predicative (or 'linking') form of 'to be,' which is why predication here is a related issue.

Metaphysics,[16] in which the O.O.M. is discussed at great length so that Aristotle can try to demolish it. There's several shelves worth of context here that we can mostly skip.[17] What's strange (for reasons that are upcoming) is that Aristotle's best-known argument against Plato's Theory of Forms is virtually textbook Zeno. This argument, which is usually called the Third Man, is in effect a divergent-VIR-type reductio on the O.O.M. After observing that both individual men and the Form Man obviously share some predicable quality or attribute, Aristotle points out that there must then be yet another metaphysical Form—say, Man′—that comprises this common attribute, which entails still another Form, Man″, to comprise the predicable commonality between Man′ and [Man + men], & c. & c. ad inf.

Whether or not the Third Man strikes you as a valid refutation of the O.O.M., you may well have already noticed that Plato's Theory of Forms[18] has problems of its own, like for example a conspicuous goofiness when the O.O.M. is applied to certain predicates—is there an ideal Form of left-handedness? of stupidity? of shit? Note, however, that Plato's theory has a lot more power and plausibility when applied to any kind of system that depends on formal relations between abstractions.

[16] **IYI** in Book I, Chs. 6 & 9. Plus of course this book's title is where the term originated; all it originally meant was that it was Aristotle's next treatise after the *Physics*.

[17] One or two factoids. Plato, né Aristocles, is c. 427–347 BCE; Aristotle is 384–322 BCE (compare Socrates at c. 470–399 and Zeno at c. 490–435). Aristotle was a former star pupil in Plato's Academy, the motto over the front door of which happened to be LET NO ONE WHO IS IGNORANT OF GEOMETRY ENTER HERE.

[18] or at least our simplified version of it.

Like math. The conceptual move from 'five oranges' and 'five pennies' to the quantity five and the integer 5 is precisely Plato's move from 'man' and 'men' to Man. Recall, after all, Hardy's thrust in §1c: when we use an expression like '2 + 3 = 5,' what we're expressing is a general truth whose generality depends on the total abstractness of the terms involved; we are really saying that two of anything plus three of anything will equal five of anything.

Except we never actually say that. Instead we talk about the *number* 2 and the *number* 5, and about relations between these numbers. It's worth it—again—to point out that this could be just a semantic move, or it could be a metaphysical one, or both. And worth it to recall both §1d's thing about the predicative v. nominative senses of 'length' and 'gram,' and the different types of existence-claims involved in 'I see nothing on the road' and 'Man is by nature curious' and 'It is raining'; and then to consider, carefully, the existence-claims we're committing to when we talk about numbers. Is '5' just some kind of conceptual shorthand for all the actual quintuples in the world?[19] It's pretty apparent that it's not, or at least that this isn't all '5' is, since there are lots of things about 5 (e.g. that 5 is prime, that 5's square root is 2.236 . . .) that don't have anything to do with real-world quintuples but do have to do with a certain kind of entity called numbers and with their qualities and relations. Numbers' real, if strange,

[19] **IYI** This is a bit esoteric, but to head off possible objections to what follows: Yes, in some sense, if '5' is understood as referring to or picking out the set of all quintuples, then by Peano's Postulates the above is exactly what '5' is—although both 'set' and Peano's Postulates are themselves Cantor-dependent, so we are literally 2000 years ahead of ourselves.

existence is further suggested by the way many of these qualities and relations—such as for example that $\sqrt{5}$ cannot be expressed as either a finite decimal or a ratio of integers—seem like they really are *discovered* rather than made up or proposed and then defended. Most of us would be inclined to say that $\sqrt{5}$ is an irrational number even if nobody ever actually proves that it is—or at least it turns out that to say anything else is to be committed to a very complex and strange-looking theory of what numbers are. The whole issue here is of course incredibly hairy (which is one reason we're talking about it only in little contextual chunks), because not only is the question abstract but everything it's concerned with is an abstraction—existence, reality, number . . .. Although consider too for a moment how many levels of abstraction are involved in math itself. In arithmetic there's the abstraction of number; and then there's algebra, with a variable being a further-abstracted symbol for some number(s) and a function being a precise but abstract relation between domains of variables; and then of course there's college math's derivatives and integrals of functions, and then integral equations involving unknown functions, and differential equations' families of functions, and complex functions (which are functions of functions), and definite integrals calculated as the difference between two integrals; and so on up through topology and tensor analysis and complex numbers and the complex plane and complex conjugates of matrices, etc. etc., the whole enterprise becoming such a towering baklava of abstractions and abstractions of abstractions that you pretty much have to pretend that everything you're manipulating is an actual, tangible thing or else you get so abstracted that you can't even sharpen your pencil, much less do any math.

The most relevant points with respect to all this are that the question of mathematical entities' ultimate reality is not just vexed but controversial, and that it was actually G. F. L. P. Cantor's theories of ∞ that brought this controversy to a head in modern math. And that in this controversy, mathematicians who tend to regard mathematical quantities and relations as metaphysically real are called Platonists,[20] and at least now it's clear why, and the term can be thrown around later.

§2b. The first really serious non-Platonist is Aristotle. What's odd and ironic about the Zenoish VIR Aristotle runs against Plato's metaphysics, however, is that Aristotle's is also the first and most important Greek attempt to refute Zeno's Paradoxes. This is mainly in Books III, VI, and VIII of the *Physics* and Book IX of the *Metaphysics*, whose discussions of Zeno will end up having a pernicious effect on the way math handles ∞ for the next two millennia. At the same time, Aristotle does manage to articulate the root difficulties in at least some of Zeno's Paradoxes, as well as to pose clearly and for

[20] **IYI** See for example this classic Platonist statement by C. Hermite (1822–1901, big number-theorist):

> I believe that the numbers and functions of analysis are not the arbitrary product of our spirits: I believe that they exist outside of us with the same character of necessity as the objects of objective reality; and we discover them and study them as do the physicists, chemists and zoologists.

As will probably emerge in various contexts below, B. P. Bolzano, R. Dedekind, and K. Gödel are all Platonists, and G. F. L. P. Cantor is at least a closet Platonist.

the first time some really vital ∞-related questions that nobody until the 1800s will even try to answer in a rigorous way, viz.: 'What exactly does it mean to say that something is infinite?' and 'Of what sort of thing can we even coherently ask whether it's infinite or not?'

W/r/t these central questions, you might recall Aristotle's famous predilection for dividing and classifying—he literally put the 'analytic' in analytic philosophy. See for instance this snippet from *Physics* VI's discussion of the Dichotomy: "For there are two senses in which length and time and generally anything continuous are called 'infinite': they are called so either in respect of divisibility or in respect of their extremities [= size]," which happens to be the first time anyone had ever pointed out that there's more than one sense to 'infinite'. Aristotle mainly wants to distinguish between a strong or quantitative sense, one meaning literally infinite size or length or duration, and a weaker sense comprising the infinite divisibility of a finite length. The really crucial distinction, he claims, involves time: "So while a thing in a finite time cannot come in contact with things quantitatively infinite, it can come in contact with things infinite in respect of divisibility: for in this sense the time itself is also infinite."

Both the above quotations are from one of Aristotle's two main arguments against Zeno's Dichotomy as schematized on pp. 49–50. The target of this particular argument is premise (3)'s 'in a finite amount of time'. Aristotle's thrust is that if Zeno gets to represent the interval AB as the sum of an infinite number of subintervals, the allotted time it takes to traverse AB should be represented the same way—say like $\frac{t}{2}$ to get to $\frac{AB}{2}$, $\frac{t}{4}$ to get to $\frac{AB}{4}$, $\frac{t}{8}$ to get to $\frac{AB}{8}$, etc. This argument isn't all that helpful, though, since having an infinite amount of time to cross the street is no less contradictory of our actual ten-second street-crossing experiences than the original Dichotomy itself. Plus it's easy to construct versions of the Z.P. that don't explicitly require action or elapsed time. (For example, imagine a pie the first piece of which = half the whole pie and the next piece = half the first piece and dot dot dot ad inf.: is there a last piece of pie or not?) The point: Counterarguments about sequential time or subintervals or even actual human movements will always end up impoverishing the Dichotomy and failing to state the real difficulties involved. Because Zeno can amend his presentation and simply say that being at A and then being at B requires you to occupy the infinitely many points corresponding to the sequence $\frac{AB}{2}, \frac{AB}{4}, \frac{AB}{8}, \ldots \frac{AB}{2^n}, \ldots$, or, worse, that your ever really arriving at B entails your having *already occupied* an infinite sequence of points. And this seems quite clearly to contradict the idea of an infinite sequence: if '∞' really means 'without end,' then an infinite sequence is one where, however many terms are taken, there are still others that remain to be taken. Meaning forget street-crossing or nose-touching: Zeno can run the whole cruncher in terms of abstract sequences and the fact that there is something inherently contradictory or paradoxical in the idea of an infinite sequence ever being completed.

It is against this second, more abstract and damaging version of the Dichotomy that Aristotle advances his more influential argument. This one depends on the semantics of 'infinite,' too, but it's different, and focuses on the same sorts of predicative questions that arise in the O.O.M. and Zeno's

Location Paradox. In both the *Physics* and the *Metaphysics*, Aristotle draws a distinction between two different things we can really mean when we use 'to be' + ∞ in a predicative sentence like 'There are an infinite number of points that must be occupied between A and B.' The distinction is only superficially grammatical; it's really a metaphysical one between two radically different existence-claims implicit in the sentence's 'are,' which apparently the Dichotomy depends on our not seeing. The distinction is between *actuality* and *potentiality* as predicable qualities; and Aristotle's general argument is that ∞ is a special type of thing that exists potentially but not actually, and that the word 'infinite' needs to be predicated of things accordingly, as the Dichotomy's confusion demonstrates. Specifically, Aristotle claims that no spatial extension (e.g. the intercurb interval AB) is 'actually infinite,' but that all such extensions are 'potentially infinite' in the sense of being infinitely divisible.

This all gets extremely involved and complex, of course—entire careers are spent noodling over Aristotle's definitions. Suffice here to say that the actual-v.-potential-existence-of-∞ issue is vital to our overall Story but admittedly tough to get a handle on. It doesn't help matters that Aristotle's own explanations and examples—

> [I]t is as that which is building is to that which is capable of building, and the waking to the sleeping, and that which is seeing to that which has its eyes shut but has sight. Let actuality be defined by one member of this antithesis, and the potential by the other. But all things are not said in the *same sense* to exist actually, but only by analogy—as A is in B or to B, C is in D or to D; for some are as movement to potency, and the others as substance to some sort of matter . . .

—are not exactly marvels of perspicuity. What he means by 'potential' is emphatically not the sort of potentiality by which a girl is potentially a woman or an acorn an oak. It's rather more like the strange and abstract sort of potentiality by which a perfect copy of Michelangelo's *Pieta*[21] potentially exists in a block of untouched marble. Or, ∞-wise, the way anything that occurs cyclically (or, in A.'s term, "successively")—like say its being 6:54 A.M., which happens every day like clockwork—is for Aristotle potentially infinite in the sense that an endless periodic recurrence of its being 6:54 A.M. is possible, whereas the set of all 6:54 A.M.s cannot be actually infinite because the 6:54s are never all going to coexist; the periodic cycle is never going to be "complete[d]."[22]

You can probably see how all this is going to play out w/r/t the Dichotomy. Again, though, it's a little tricky. The statue and 6:54 analogies won't quite work here. Yes, the interval AB and/or the set of all subintervals or points between A and B is not 'actually infinite' but only 'potentially infinite'; but here the sense in which Aristotle means 'AB is potentially infinite' is closer to the idea of, say, infinite precision in measurement. Which can be illustrated thusly. My eldest niece's

[21] as well, of course, as every other statue ever done, or thought of, or even not thought of. . . .

[22] **IYI** Like so much of Aristotle, this is not immediately clear. The thing here is that 'coexist' basically means 'all exist at the same time,' which the 23-hour-and-59-minute gap between each recurrence of 6:54 A.M. (these gaps being packed into the very definition of '6:54 A.M.') renders impossible. In essence, this Succession → Noncompletion thing is also Aristotle's argument for why capital-T Time is potentially but not actually infinite, which in turn preempts certain Lamp-type paradoxes about eternity and first and last moments.

current height, which is 38.5″, can be fixed more precisely at 38.53″; and with a more controlled environment and sophisticated equipment it could obviously be ascertained more and more exactly, to the 3rd, 11th, *n*th decimal place, with *n* being, potentially, ∞—but only *potentially* ∞, because in the real world there's obviously never going to be any way to achieve true infinite precision, even though 'in principle' it's possible. In pretty much just this way, for Aristotle AB is 'in principle' infinitely divisible, though this infinite division can never actually be performed in the real world.

(**IYI** Final bit of complication: For the most part, what Aristotle calls "Number" (meaning mathematical quantities in general) apparently is potentially infinite not in the way measurement is potentially infinite but in the way the set of all 6:54 A.M.s is potentially infinite. For instance, the set of all integers is potentially infinite in the sense that there is no largest integer ("In the direction of largeness it is always possible to think of a larger number"); but it is not actually infinite because the set doesn't exist as one completed entity. In other words, numbers for Aristotle compose a successive continuum: there are infinitely many but they never coexist ("One thing can be taken after another endlessly").)

As a refutation of Zeno's Dichotomy, the potential-∞-v.-actual-∞ distinction isn't all that persuasive—evidently not even to Aristotle, whose own Third Man regressus looks like it could be dismissed as only potentially infinite. But the distinction ends up being terribly important for the theory and practice of math. In brief, relegating ∞ to the status of potentiality allowed Western math either to discount infinite quantities or to justify their use, or sometimes both, depending on the agenda. The whole thing is very weird. On the one hand,

Aristotle's argument lent credence to the Greeks' rejections of ∞ and of the 'reality' of infinite series, and was a major reason why they didn't develop what we now know as calculus. On the other hand, granting infinite quantities at least an abstract or theoretical existence allowed some Greek mathematicians to use them in techniques that were extraordinarily close to being differential and integral calculus—so close that in retrospect it's amazing that it took 1700 years for actual calc to be invented. But, back on the first hand, a big reason it *did* take 1700 years was the metaphysical shadowland Aristotle's potentiality concept had banished ∞ to, which served to legitimate math's allergy to a concept it couldn't really ever handle anyway.

Except—either back on the second hand or now on a third hand[23]—when G. W. Leibniz and I. Newton now really do introduce the calculus around 1700, it's essentially Aristotle's metaphysics that justifies their deployment of infinitesimals, e.g. dx in the infamous $\frac{f(x + dx) - f(x)}{dx}$ of freshman math. Please either recall or be informed that an infinitesimal quantity is somehow both close enough to 0 to be ignorable in addition—i.e., $x + dx = x$—and distant enough from 0 to serve as a divisor in derivations like the above. Again very briefly, treating infinitesimals as potentially/theoretically existent quantities let mathematicians use them in calculations that had extraordinary real-world applications, since they were able to abstract and describe just the kinds of smooth continuous phenomena the world comprised. These infinitesimals turn out to be a very big deal. Without them you can

[23] **IYI** and to anticipate some stuff that will be discussed at length in §4.

end up in crevasses like the .999 . . . = 1 thing we glanced at in §1c. As was then promised, the quickest way out of that one is to let x stand not for .999 . . . but for the quantity 1 minus some infinitesimal, which let's call $\frac{1}{\infty}$, such that $1 > (1 - \frac{1}{\infty}) > .999 \ldots$. Then you can run the same operations as before:

$$10x = 10 - (\tfrac{10}{\infty})$$

$$-\ x, \text{ which} = 1 - (\tfrac{1}{\infty}),$$

yields $9x = 9 - (\frac{9}{\infty})$,

in which case x still comes out to $1 - (\frac{1}{\infty})$ and there's no nasty confabulation with 1.0.

Except of course the question is whether it makes metaphysical or mathematical sense to posit the existence, whether actual or potential, of some quantity that is <1 but still exceeds the infinite decimal .999 . . .. The issue is doubly abstract, since not only is .999 . . . not a real-world-type quantity, it's something we cannot really conceive of even as a mathematical entity; whatever relationship there is between .999 . . . and $(1 - \frac{1}{\infty})$ exists[24] out past the *n*th decimal, a place no one and nothing can ever get to, not even in theory. So it's not clear whether we're just trading one kind of paradoxical crevasse for another. This is yet another type of question that is totally vexed before Weierstrass, Dedekind, and Cantor weigh in in the 1800s.

Whatever you might think of Aristotle's potential-type ontology for ∞, notice that he was at least right to home in on

[24] so to speak.

words like 'point' and 'exist' in the nonpredicative sentence 'There are [= exist] an infinite number of intermediate points between A and B.' Just as in Zeno's Location Paradox, there's obviously some semantic shiftiness going on here. In the revised Dichotomy, the shiftiness lies in the implied correspondence between an abstract mathematical entity—here an infinite geometric series—and actual physical space. It's not clear that 'exist' is the more vulnerable target, though; there's a rather more obvious ambiguity in the semantics of 'point'. If A and B are the two sides of a real-world street, then the noun phrase 'the infinite number of points between A and B' is using 'point' to denote a precise location in physical space. But in the noun phrase 'the infinite number of intermediate points designated by $\frac{AB}{2^n}$,' 'point' is referring to a mathematical abstraction, a dimensionless entity with 'position but no magnitude'. To save several pages of noodling that you can do in your own spare time,[25] we'll simply observe here that traversing an infinite number of dimensionless mathematical points is not obviously paradoxical in the way that traversing an infinite number of physical-space points is. In this respect, Zeno's argument can look rather like §1's three-men-at-motel brainteaser: the translation of an essentially mathematical situation into natural language somehow lulls us into forgetting that regular words can have vastly different senses and referents. Note, one more time, that this is exactly what the abstract

[25] Just as a sort of prompt: It's not trivial to observe that the ancient Greeks had no true conception of a dimensionless point, something with zero extension, since they didn't have 0. And maybe thus that in a sense the Dichotomy was just a symptom of the Greeks' real problem, which was their attempt to do abstract math using only concrete quantities.

symbolism and schemata of pure math are designed to avoid, and why technical math definitions are often so numbingly dense and complex. You want no room for ambiguity or equivocation. Mathematics, like child-measurement, is an enterprise consecrated to the ideal of precision.

Which all sounds very nice, except it turns out that there is also immense ambiguity—formal, logical, metaphysical—in many of the basic terms and concepts of math itself. In fact the more fundamental the math concept, the more difficult it usually is to define. This is itself a characteristic of formal systems. Most of math's definitions are built up out of other definitions; it's the really root stuff that has to be defined from scratch. Hopefully, and for reasons that have already been discussed, that scratch will have something to do with the world we all really live in.

§2c. Back for a moment to the Zeno-and-semantics-of-'point' thing. The relation between a mathematical entity (e.g. a series, a geometric point) and actual physical space is also the relation of the discrete to the continuous. Think of a flagstone path v. a shiny smooth black asphalt road. Since what the Dichotomy tries to do is break a continuous physical process down into an infinite series of discrete steps, it can be seen as history's first-ever attempt to represent continuity mathematically. It doesn't matter that Zeno was actually trying to show that continuity was impossible; he was still the first. He was also the first to recognize[26] that there is more

[26] **IYI** the first in practice, anyway—Book VI of the *Physics* gets credit for making it explicit.

than one species of ∞. The *to apeiron* of Greek cosmology is pure extension, infinite size; and the integral series 1, 2, 3, . . . ascends and recedes toward this same kind of Big ∞. Whereas on the other hand Zeno's Little ∞ appears to be nested amid and between ordinary integers. Which latter is naturally hard to conceive.

It turns out that the most perspicuous way to represent these two different kinds of ∞ is with the good old Number Line, yet another feature of the ordinary 2nd-grade classroom.[27] The Number Line is also another bequest from the Greeks, who you'll recall treated numbers and geometrical shapes as pretty much the same thing. (Euclid, for instance, rejected any piece of mathematical reasoning that could not be "constructed," meaning demonstrated geometrically.[28]) The thing to appreciate about the humble Number Line's marriage of math and geometry is that it's also the perfect union of form and content. Because each number corresponds to a point, and because the Number Line both comprises all the points and determines their order, numbers can be wholly defined by their place on the N.L. relative to other numbers' places. As in, 5 is the integer immediately to the right of 4 and to the left of 6, and to say that $5 + 2 = 7$ is to say that 7 is two positions to the right of 5—that is, the mathematical 'distance' between unequal numbers can be represented and even calculated pictorially. Even without zero or

[27] **IYI** This is the thing that usually ran above the blackboard (or along the top of the back wall in classrooms that had U.S. Presidents' portraits running over the board) and looked kind of like a thermometer on its side.

[28] **IYI** A completely different criterion of 'constructibility' for theorems will become important much later in §6f.

negative integers,[29] and with 'point' being rather fuzzily defined by Euclid as "that which has no part," the Number Line is an immensely powerful tool. It also happens to be the perfect schematization of a *continuum*, meaning 'an entity or substance whose structure or distribution is continuous and unbroken,' and as such the N.L. embodies perfectly the antinomy of continuity that Zeno proposed and no one 'til R. Dedekind could solve. For on the N.L. the following are both true: (1) Every point is next to another point; (2) Between any two points there is always another point.

Even though everyone knows what it looks like, the Number Line[30] is reproduced here, starting at the 0 the Greeks didn't have because for now it doesn't matter:

$$\underbrace{0 \quad 1}_{\infty_1} \quad 2 \quad 3 \quad 4 \quad 5 \quad 6 \quad \cdots \quad n \quad \cdots \longrightarrow \infty_2$$

[29] which latter the Greeks didn't have either.

[30] For sexy technical reasons that we'll get to, the Number Line is more properly called the Real Line if it also maps the irrational numbers. Meaning 'Real Line' as in all the real numbers. Note, by the way, that another term in mathematics for both the set of all real numbers and the Real Line happens to be: *the Continuum.*

A certain amount of all this will probably get mentioned in the text, but at some point the N.L.-v.-R.L. thing has to get nutshelled, and this is as good a place as any. Be apprised that the math-metaphysics of both kinds of line are heavy indeed. They share three crucial features; the R.L. alone has a fourth. Both types of line are by definition infinitely extended; they are both infinitely dense (= between any two points there's always another); and they are both 'successive,' or 'ordered' (which basically means that for any point n, $(n - 1) < n < (n + 1)$). The Real Line alone has the quality of being *continuous,* which here means it's got no gaps or holes in it. Note for later that it's the R.L.'s continuousness in this sense that ends up being the real crevasse for modern math. As mentioned in the main text just above, though, Zeno's

If the Big ∞, the infinity of extension, lies at the endless right of the Number Line, the Little ∞ that Zeno exploits lies in the totally finite-looking interval between 0 and 1, which interval he reveals as containing an infinite number of intermediate points, viz. the sequence $\frac{1}{2}, \frac{1}{4}, \frac{1}{8}, \ldots$. What's more (so to speak), it's clear that this infinity of $\frac{1}{2^n}$ does not actually exhaust the points between 0 and 1, since it leaves out not only convergent infinite sequences like $\frac{1}{6}, \frac{1}{10}, \frac{1}{18}, \frac{1}{34}, \ldots$, $\frac{1}{12}, \frac{1}{36}, \frac{1}{72}, \frac{1}{120}, \ldots$, etc., but the whole other infinite set of fractions $\frac{1}{x}$ where x is an odd number. And when you consider that each of these latter fractions will correspond to its own infinite geometric sequence via the expansion of $\frac{1}{x^n}$—e.g. $\frac{1}{3}, \frac{1}{9}, \frac{1}{27}, \frac{1}{81}, \ldots, \frac{1}{5}, \frac{1}{25}, \frac{1}{125}, \frac{1}{15625} \ldots$, etc.—it appears that the finite N.L. interval 0–1 actually houses an infinity of infinities. Which is, to put it mildly, both metaphysically puzzling and mathematically ambiguous—like would this be ∞^2, or ∞^∞, or what?

Dichotomy requires nothing more than N.L.-grade density to create the paradox. And this is why the Dichotomy can so easily be recast to eliminate time/motion: its VIR involves traversing not real-world space but just the interval 0–1 on the Number Line. It is particularly this second, dense, inter-number type of ∞ that Aristotle wants to dismiss as merely 'potential'.

Finally, please N.B. that in certain places between here and §6 we're going to speak as if the Number Line and the Real Line were the same thing, or as if the N.L. could also map irrationals. This will be for complicated reasons involving the translation of technical proofs into natural language, none of which are your problem except maybe to just keep the Real Line's special status in mind for several §s until it becomes important.

Except it gets worse, or better. Because all the prenominate numbers are rational. You probably already know that the adjective here derives from 'ratio' and that the phrase 'the rational numbers' refers to all those numbers expressible either as integers or as ratios of two integers (that is, as fractions). This is just review, but it's important. The discovery that not all numbers are rational was at least as hard on the Greek worldview as Zeno's Paradoxes. And it was particularly upsetting to the Divine Brotherhood of Pythagoras. Recall the Pythagorean convictions that everything is a mathematical quantity or ratio and that nothing infinite can really exist in the world (since (*peras* → form) is what enables existence in the first place).

Then recall the Pythagorean Theorem. As mentioned, an interesting bit of trivia is that the D.B.P. were not the true discoverers of this theorem; it actually shows up in Old Babylonian tablets as early as 2000 BCE. One reason it's called the Pythagorean Theorem is that it enabled the D.B.P.'s discovery of 'incommensurable magnitudes,' a.k.a. irrational numbers or surds.[31] These numbers, which turn out to be inexpressible as finite quantities, were so lethal to Pythagorean metaphysics that their discovery became sort of the Greek version of Watergate. You will remember from childhood that the Pythagorean Theorem causes no problems with figures like the 3-4-5 right triangle of Intro Geometry, wherein the sum of the squares of 3 and 4 is a number whose own square root is rational. Understand, though,

[31] **IYI** The latter was Dr. Goris's preferred term because he maintained that it was so much more fun to say. If you said 'irrational number' he'd pretend he couldn't hear you, which if you know the etymology of 'surd' you'll see was itself a kind of in-joke.

that this 'squares of' stuff was literal for the Greeks. That is, in a 3-4-5 triangle they treated each leg as the side of a square—

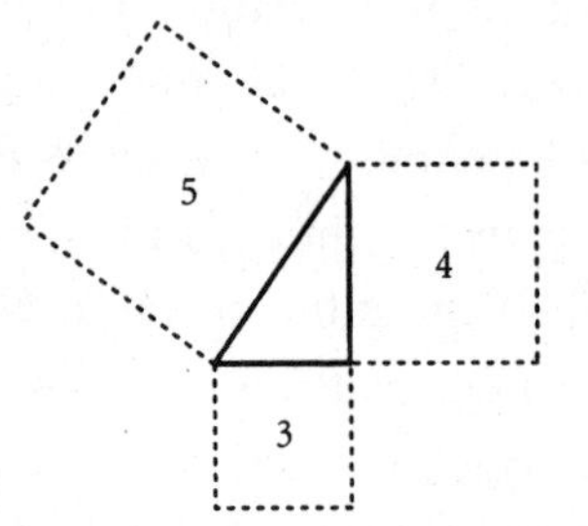

—and then added up the areas of the squares. There are two reasons this is noteworthy. The first was mentioned someplace above: while we now toss exponents and radicals around *in abstractus*, math problems for the Greeks were always formulated, and solved, geometrically. A rational number was a literal ratio of two line-lengths; squaring something was constructing a square and taking its area. The second reason is that by most accounts it was a plain old humble square that started all the trouble. Consider specifically the familiar Unit Square, with sides equal to 1, and even more specifically the isosceles right triangle whose hypotenuse is the Unit Square's diagonal:

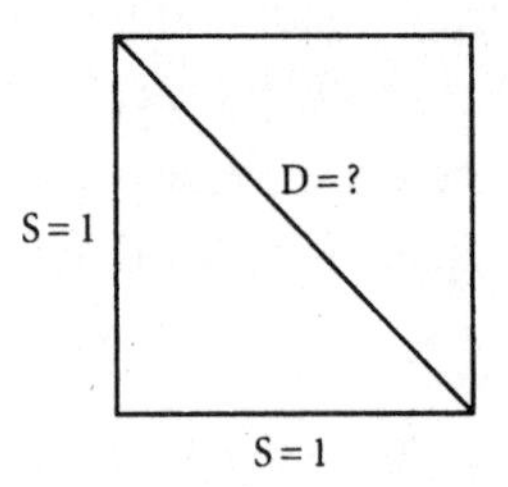

What the D.B.P. realized (probably through actual and increasingly frantic measurements) is that no matter how small a unit of measure is used, the side of a Unit Square is *incommensurable* with the diagonal. Meaning there is no rational number $\frac{p}{q}$ such that $\frac{D}{S} = \frac{p}{q}$. The quantity $\frac{D}{S}$ was something the D.B.P. eventually called *arratos*, 'the not-having-a-ratio,' or—since *logos* could mean both word and proportion—*alogos*, which thus meant both 'the nonproportional' and 'the unsayable'.

The actual demonstration of the incommensurability of $\frac{D}{S}$ is another famous instance of reductio proof, and an especially nice one because it's very simple and requires only jr.-high math. So here it is. First, for reductio purposes, assume that D and S are commensurable. This means that $\frac{D}{S}$ equals some ratio $\frac{p}{q}$ where p and q are integers with no common factor greater than 1. We know by the Pythagorean Theorem that $D^2 = S^2 + S^2$, or $D^2 = 2S^2$, which if $\frac{D}{S} = \frac{p}{q}$ means that $p^2 = 2q^2$. We know further that the square of any odd number is going to be odd and the square of any even number will be even (feel free to test these out). Plus we know that anything times 2 will obviously be even. This is all the ordnance we need. By LEM, either p is odd or p is even. If (1) p is odd, there's an immediate contradiction, since $2q^2$ has got to be even. But if (2) p is even, that means it's equal to some number times 2, say $2r$, so plugging this equivalence back into the original $p^2 = 2q^2$ yields $4r^2 = 2q^2$, which reduces to $2r^2 = q^2$, which means q^2 is even, which means q is even, which means both p and q are even, in which case they have a common factor greater than 1, which is again a contradiction. (1) yields a contradiction; (2) yields a contradiction; there is no (3). So D and S are incommensurable. End proof.[32]

The fact that rational numbers couldn't express something as quotidian as the diagonal of a square—not to mention other easily-constructed hypotenoid irrationals like $\sqrt{5}$, $\sqrt{8}$, etc.—was obviously destabilizing to the Pythagoreans' whole cosmogony. The coup de grâce was apparently the discovery that their beloved Golden Mean was itself irrational, working out to $\frac{1}{2}(1 + \sqrt{5})$ or 1.618034 . . .. There's all sorts of lurid apocrypha about the ends the D.B.P. supposedly went to to keep the existence[33] of irrationals secret, which we can skip because much more important, historically and mathematically, are surds themselves. They're important for at least three reasons. (1) Mathematically, irrational numbers are a direct consequence of abstraction. They're a whole level up from 5 oranges or $\frac{1}{2}$ a pie; you don't encounter irrationals until you start generating abstract theorems like the P.T. And please note that they're really only a crevasse for pure math. The Egyptians et al. had run into irrationals in surveying and engineering, but because they cared only about practical applications they had no problem with treating a quantity like $\sqrt{2}$ as 1.4 or $\frac{7}{5}$. (2) Surds' discovery marked the first real divergence of math and geometry, the former now able to manufacture numbers that geometers couldn't actually

[32] **IYI** This also of course functions as a proof that $\sqrt{2}$ is irrational, which was the way Dr. G. first presented it in class.

[33] so to speak.

measure. (3) It turns out that irrationals, just like Zeno's $\frac{1}{2^n}$'s, are a consequence of trying to express and explain continuity w/r/t the Number Line. Irrational numbers are the reason why the Number Line isn't technically continuous. Like the Dichotomy's VIR, surds represent gaps or holes in the N.L., interstices through which the limitless chaos of ∞ could enter and mess with the tidiness of Attic math.

And it's not just a Greek problem. Because the big thing about irrational numbers is that they can't be represented by fractions; and yet if you try to express irrationals in decimal notation,[34] then the sequence of digits after the decimal will be neither *terminal* (as in the rational decimals 2.0, 5.74) nor *periodic* (which means repeating in some kind of pattern, as in the rationals $0.333\ldots = \frac{1}{3}$, $1.181818\ldots = \frac{13}{11}$, etc.).[35]

[34] **IYI** itself an invention of the sixteenth century.

[35] It's also important to keep in mind that decimals are really just *numerals*, meaning representations of numbers rather than numbers themselves. And that decimals also happen to be representations of convergent series, with e.g. '0.999 . . . ' being equivalent to $\frac{0}{10^0} + \frac{9}{10^1} + \frac{9}{10^2} + \frac{9}{10^3} + \cdots$. If you're able to see why the sum of this particular infinite series is 1.0, it's probably going to occur to you that the above-mentioned 0.999 . . . = 1.0 paradox isn't really a paradox at all but merely a consequence of the fact that there's always more than one way to represent any given number in decimal notation. In this case, the quantity 1 can be expressed either as '1.000 . . . ' or as '0.999 . . . '. Both representations are valid, although you need a certain amount of college math to see why. (**IYI** Again, if you can keep stuff in your head for long intervals and many pages, know now that in §§ 6 and 7, G. Cantor is going to make ingenious use of this technical equivalence between 1.0 and .999 . . . in a couple of his most famous proofs.)

Meaning that, for example, the decimal expression of $\sqrt{3}$ can be carried to 1.732, or 1.73205, or 1.7320508, or literally as long as you like . . . and longer. Meaning in turn that a certain definite point on the Number Line—viz. the point corresponding to that interval which, multiplied by itself, corresponds to the integer-point 3—cannot be named or expressed finitely.[36]

The finite interval 0–1 on the Number Line is thus even more inconceivably crowded. There's not only an infinite number of infinite sequences of fractions, but also an infinite number of surds,[37] each of which is itself numerically inexpressible except as an infinite sequence of nonperiodic decimals. Let's pause to consider the vertiginous levels of abstraction involved here. If the human CPU cannot apprehend or even really conceive of ∞, it is now apparently being asked to countenance an infinity of ∞s, an infinite number of individual members of which are themselves not finitely expressible, all in an interval so finite- and innocent-looking we use it in little kids' classrooms. All of which is just resoundingly weird.

There are, of course, as many ways to handle this weirdness as there are connotations for 'handle'. The Greeks,* for instance, simply refused to treat irrationals as numbers. They either categorized them as purely geometric lengths/areas and never used them in their math per se, or they literally rationalized the use of surds by futzing with their arithmetic

[36] **IYI** Not numerically, anyway. Other ways to say the same thing: In the dental-sounding nomenclature of high-school math, the root of 3 is not fully extractable; in graphical terms, a line with an irrational slope will never hit any point corresponding to a Cartesian coordinate.

[37] To see intuitively that there are (at the very least) an infinite number of irrational points between 0 and 1, consider the set of all points corresponding to $\frac{1}{n}$ where n is irrational.

(example: the Pythagoreans' eventual trick was to write 2 as $\frac{49}{25}$ so they could treat $\sqrt{2}$ as $\frac{7}{5}$).[38] If their refusal to acknowledge the existence of numbers that their own mathematical reasoning produced seems kind of bizarre, be apprised that up to the 1700s pretty much all the best mathematicians in Europe did the same thing,[39] even as the fabled Scientific Revolution was starting to produce all kinds of results that required an arithmetic of irrationals. Not until the late nineteenth century,* in fact, would anyone* come up with a rigorous theory or even definition of irrationals. The best definition would come from R. Dedekind,* while the most comprehensive treatment of real numbers' status on the Line would be G. Cantor's.

[38] **IYI** This is ultimately why Greek trig and astronomy were such a mess—they tried to quantify continuous curves and sub-curve areas with only rational numbers.

[39] Q.v. this vividly appropriate quotation from the German algebrist M. Stifel, c. 1544:

> Since, in proving geometrical figures, when rational numbers fail us irrational numbers take their place and prove exactly those things which rational numbers could not prove, we are compelled to assert that they truly are numbers. On the other hand, other considerations compel us to deny that irrational numbers are numbers at all. To wit, when we seek to subject them to [decimal representation], we find that they flee away perpetually, so that not one of them can be apprehended precisely in itself. Now, that cannot be called a true number which is of such a nature that it lacks precision. Therefore, just as an infinite number is not a number, so an irrational number is not a true number, but lies hidden in a kind of cloud of infinity.

§2d. ***UNAVOIDABLE BUT ULTIMATELY IYI-GRADE INTERPOLATION**

Skip the following few pages if you like, but the asterisks in the above ¶ tag stuff that historically speaking is not 100% true. To wit: A certain student and protégé of Plato known as Eudoxus of Cnidos (408–354 BCE) came very close indeed to providing a rigorous definition of irrationals, which Euclid then included as Definition 5 in Book V of the *Elements.* Eudoxus's definition involves geometric proportions and ratios—which is unsurprising, given that Greek math had been confronted by irrationals in the form of certain geometric proportions that couldn't be expressed as ratios. Following the D.B.P.'s debacle, these *incommensurable magnitudes* seemed to be everywhere—like consider a rectangle two of whose sides equal the diagonal of the Unit Square: how were you supposed to calculate its area? More important, how could the Greeks distinguish cases of irrational-type incommensurability from cases where you simply have different species of magnitudes that can't be compared via ratios, like a line v. an area or an area v. a 3D volume? Eudoxus was actually the first Greek who even tried to define 'ratio' mathematically—

> Magnitudes are said to be in the same ratio, the first to the second and the third to the fourth, when, if any equimultiples whatever be taken of the first and the third, and any equimultiples whatever of the second and the fourth, the former equimultiples alike exceed, are alike equal to, or are alike less than, the latter equimultiples taken in corresponding order.

—the opacity of which can be mitigated by translating some of the theorem's natural-language stuff into basic math symbolism. Eudoxus's def. here states that, given $\frac{p}{q}, \frac{r}{s}$, and the integers a and b, $\frac{p}{q} = \frac{r}{s}$ if and only if $(ap < bq) \rightarrow (ar < bs)$ and $(ap = bq) \rightarrow (ar = bs)$ and $(ap > bq) \rightarrow (ar > bs)$. This may at first look obvious or trivial[40]—see for instance how it resembles the rule about cross-multiplying fractions we all learned in 4th grade. But it really isn't trivial at all. Though Eudoxus meant it to apply only to geometric magnitudes rather than numbers per se, the definition works perfectly to identify and distinguish rational numbers from irrational numbers from immiscibly different geometric quantities, etc. Plus please notice now the way Eudoxus's definition is effectively able to operate on a whole infinite set, viz. that of all rational numbers.[41] What Eudoxus does is use arbitrary integers to specify a division[42] of the set of all rationals into two subsets: the set of all rationals for which $ap \leq bq$ and the set of all rationals for which $ap > bq$. His is thus the first theorem to be able to range, comprehensively and specifically,[43] over an entire infinite collection. In this respect it could be called the first significant result in set theory, about 2300 years before the invention of set theory.

[40] **IYI** If it doesn't, then be apprised/reminded that, by the rules of formal logic, an entailment like '$(ap < bq) \rightarrow (ar < bs)$' will be false *only* when the first term is true and the second term is false. Given this, feel free to let, say, $p = 1$, $q = 2$, $r = 2$, $s = 4$, $a = 2$, and $b = 1$, and to work these three different entailments out. You'll find that there's no case in which the first term's true and the second one's false—i.e., that $\frac{1}{2}$ really does equal $\frac{2}{4}$.

[41] **IYI** This will become way more relevant when we get to R. Dedekind's theory of real numbers in §6.

[42] **IYI** which in Dedekind's theory will be called a *cut*.

[43] the specificity lying in the choice of values for a and b.

It is also worth pointing out that there are probably no better examples in math of Russell's dictum about the caprice of intellectual fame than Eudoxus and his posthumous collaborator Archimedes (287–212 BCE). The latter, granted, is anecdotally famous for his '*Eureka!*' thing; but given our overall purposes it would be unfair not to acknowledge that he and Eudoxus more or less invented modern math, which then had to be reinvented many centuries later because nobody'd bothered to pay attention to the consequences of their results.

Probably their most important invention is known as the Exhaustion Property, which Eudoxus discovered and Archimedes refined. It was a way to calculate the areas and volumes of curved surfaces and figures, something that Greek geometry obviously had a lot of trouble with (since it's w/r/t curves that you encounter most of the problems of continuity and irrationals). Geometers before Eudoxus had had the idea of approximating the area of a curved figure by comparing it to regular polygons[44] whose areas they could calculate exactly. By way of example, see how the very largest square that can fit inside a circle functions as a crude approximation of the circle's area—

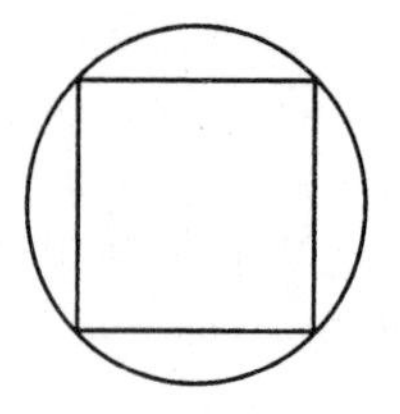

[44] **IYI** = those w/ all sides the same length.

—whereas, say, the largest octagon that can fit inside will be a slightly better approximation—

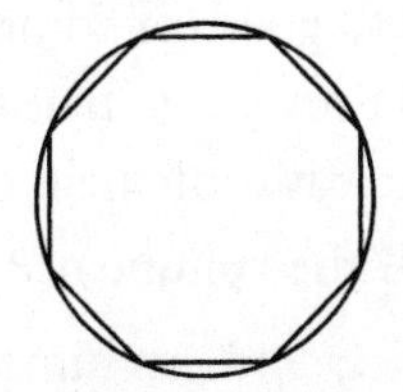

—and so on, the point being that the more sides the inscribed polygon has, the closer its area will be to the circle's own *A*. The reason the method never actually worked is that you'd need a ∞-sided polygon to nail *A* down all the way, and even if this ∞ was merely one of Aristotle's potential ∞s the Greeks were still stymied, for the same reason mentioned w/r/t the Dichotomy: they didn't have the concept of convergence-to-a-limit. Eudoxus gave math just such a concept with his introduction of the Exhaustion Property, which appears as Prop. 1 in Book X of the *Elements*:

> If from any magnitude there be subtracted a part not less than its half, and if from the remainder one again subtracts not less than its half, and if this process of subtraction is continued, ultimately there will remain a magnitude less than any preassigned magnitude of the same kind.

In modern notation, this is equivalent to saying that if p is a given magnitude and r a ratio such that $\frac{1}{2} \leq r < 1$, then the limit of $p(1 - r)^n$ is 0 as n tends to ∞—i.e., $\lim_{n \to \infty} p(1 - r)^n = 0$. This allows you to approach, arbitrarily closely, an infinite number of sides to a polygon, or an infinite number of rectangles under a curve, each side/rectangle being arbitrarily (= infinitesimally) small, and then to sum the relevant sides/areas by the inverse of the very process by which you derived them. The Method of Exhaustion is, for all intents and purposes, good old integral calculus. With it, Eudoxus was able to prove, e.g., that the ratio of any two circles' areas equals the ratio of their radii's squares, that the volume of a cone is $\frac{1}{3}$ the volume of a cylinder with the same base and height, & c.; and Archimedes' *Measurement of a Circle* uses Exhaustion to derive an unprecedentedly good approximation of π as $\frac{223}{71} < \pi < \frac{22}{7}$.

Notice also the metaphysical canniness of Exhaustion's abstract entities. Eudoxus's method of getting infinitesimally small sides/figures into equations makes no claims about the existence of infinitely tiny magnitudes. Look at the bland language of the *Elements*'s Prop. 1 above. The "less than any preassigned magnitude" is particularly clever—and strikingly similar to modern analysis's 'arbitrarily large/small'.[45] It's basically saying that, for mathematical purposes, you can reach magnitudes that are as small as you want, and work with them. It's this concern with method and results rather than ontology that makes Eudoxus and Archimedes so eerily modern-looking. The way their "magnitudes less than any preassigned magnitudes" are created and deployed in Exhaustion is pretty much identical to the way infinitesimals will be treated in early calculus.

Why, then, Europe had to wait nineteen centuries for actual calculus, differential geometry, and analysis is a very

[45] **IYI** It is, in truth, almost unbelievably close to the way A.-L. Cauchy will end up defining infinitesimals in terms of limits in order to avoid the various crevasses associated with infinitely small quantities, all of which gets hashed out below in §5.

long story that essentially bears out Russell's dictum. One cause is the same reason nobody thought to apply Exhaustion to Zeno's Dichotomy: the Greeks cared only about geometry, and nobody then thought of motion/continuity as abstractable into the geometry of the Number Line. Another reason is Rome, as in the Empire, whose sack of Syracuse and murder of Archimedes around 212 BCE brought an abrupt end to Hellenic math,[46] and whose hegemony over the next several centuries meant that a lot of the substance and momentum of Greek math was lost for a long time. The most efficient cause, though, was Aristotle, whose influence of course not only survived Rome but also reached new heights with the spread of Christianity and the Church from like 500–1300 CE. To boil it all way down, Aristotelian doctrine became Church dogma, and part of Aristotelian doctrine was the dismissal of ∞ as only potential, an abstract fiction and sower of confusion, *to apeiron*, the province of God alone, etc. This basic view predominated up to the Elizabethan era.

END INTERP.

§2e. (Continuation of §2c from the ¶ on pp. 80–81 with the interpolative asterisks in it) Here, as a sort of hors d'oeuvre, are some of the things that G. Cantor[47] eventually discovered about the nested ∞s of Zeno and Eudoxus. Discovered as in not just found out but actually proved. The Number Line is obviously infinitely long and comprises an infinity of points. Even

[46] A bit of drollery among math historians is that killing Archimedes was the only truly significant mathematical thing the Romans ever did.

[47] **IYI** with, as we'll see, some preliminary machete-work from B. P. Bolzano and R. Dedekind.

so, there are just as many points in the interval 0–1 as there are on the whole Number Line. In fact, there are as many points in the interval .0000000001–.0000000002 as there are on the whole N.L. It also turns out that there are as many points in the above micro-interval (or in one one-quadrillionth its size, if you like) as there are on a 2D plane—even if that plane is infinitely large—or in any 3D shape, or in all of infinite 3D space itself.

More, we know that there are infinitely many rational numbers on the infinite Number Line, and (courtesy of Zeno) that these rationals are so infinitely dense on the Line that for any given rational number there is literally no next rational number—that is, between any two rationals on the N.L. you can always find a third one. Of which fact here's a brief demo. Take any two different rationals p and q. Since they're different, $p \neq q$, which means one's bigger than the other. Say it's $p > q$. This means that on the Number Line there's at least some measurable distance, no matter how small, between q and p. Take that distance, divide it by some number (2 is easiest), and add the quotient to the smaller number q. You now have a new rational number, $q + (\frac{p-q}{2})$, between p and q. And since the number just of plain integers by which you can divide $(p - q)$ before adding the quotient to q is infinite, there are actually an infinity of rational points between any p and q. Let that sink in a moment, and then be apprised that even given the infinite density of the infinite number of rationals on the Number Line, you can prove that the total percentage of N.L.-space taken up by all the infinitely infinite rational numbers is: none. As in 0, nil, zip. The technical version of the proof is Cantor's, and notice how Eudoxian-Exhaustive in spirit it is, even in the following natural-language form, which requires a little creative visualization.

Imagine you can see the whole Number Line and every one of the infinite individual points it comprises. Imagine you want a quick and easy way to distinguish those points corresponding to rational numbers from the ones corresponding to irrationals. What you're going to do is ID the rational points by draping a bright-red hankie[48] over each one; that way they'll stand out. Since geometric points are technically dimensionless, we don't know what they look like, but what we do know is that it's not going to take a very big red hankie to cover one. The red hankie here can in truth be arbitrarily small, like say .00000001 units, or half that size, or half that half, . . ., etc. Actually, even the smallest hankie is going to be unnecessarily large, but for our purposes we can say that the hankie is basically infinitesimally small—call such a size ϕ. So a hankie of size ϕ covers the N.L.'s first rational point. Then, because of course the hankie can be as small as we want, let's say you use only a $\frac{\phi}{2}$-size hankie to drape over the next rational point. And say you go on like that, with the size of each red hankie used being exactly $\frac{1}{2}$ that of the previous one, for all the rational numbers, until they're all draped and covered. Now, to figure out the total percentage of space all the rational points take up on the Number Line, all you have to do is add up the sizes of all the red hankies. Of course, there are infinitely many hankies, but size-wise they translate into the terms in an infinite series, specifically the Zeno-esque geometric series

[48] **IYI** There are all sorts of different procedures and objects to illustrate this proof with. As it happens, Dr. Goris used always to carry, blow/mop with, and deploy illustratively a large red pocket-handkerchief, which for over 25 years of classes he referred to as the Hankie of Death.

$\frac{1}{2^0} + \frac{1}{2^1} + \frac{1}{2^2} + \frac{1}{2^3} + \frac{1}{2^4} + \cdots$; and, given the good old $\frac{a}{(1 - r)}$ formula for summing such a series, the sum-size of all the infinite hankies ends up being 2ϕ. But ϕ is infinitesimally small, with infinitesimals being (as was mentioned in §2b) so incredibly close to 0 that anything times an infinitesimal is also an infinitesimal, which means that 2ϕ is also infinitesimally small, which means that all the infinite rational numbers combined take up only an infinitesimally small portion of the N.L.—which is to say basically none at all[49]—which is in turn to say that the vast, vast bulk of the points on any kind of continuous line will correspond to irrational numbers, and thus that while the aforementioned Real Line really is a line, the all-rational Number Line, infinitely dense though it appears to be, is actually 99.999 . . . % empty space, rather like DQ ice cream or the universe itself.

Let's each pause privately for a moment to try to imagine what the inside of Professor G. F. L. P. Cantor's head might look like as he's proving stuff like this.

A canny reader here may object that there's some kind of Zenoid sleight of hand going on in the above proof, and might ask why a similar hankie procedure and series couldn't be applied to the irrational numbers to quote-unquote prove that the total % of Line-space taken up by the irrationals is also 2ϕ. The reason such a proof can't work is that, no matter how infinitely or even ∞^{∞}ly many red hankies you drape, there will always be more irrational numbers than hankies. Always. Cantor proved this, too.

[49] **IYI** You can, as a matter of fact, prove mathematically that the probability of somebody ever hitting a rational point with a random finger or dart on the N.L. or a proton fired randomly at it or whatever is: 0%.

§3a. It becomes appropriate at this point to bag all pretense of narrative continuity and to whip through several centuries schematically in a kind of timeline that goes from let's say 476 CE (fall of Rome) to the 1660s (foreplay to calculus). With the timeline's salients obviously being stuff related to ∞ and/or to the overall situation in math when Dedekind and Cantor enter the picture. And with the disadvantages of sketchy abstraction being somewhat offset by the advantages of compression, since at least one of us is starting to figure out that overall space is going to be a concern. The project here in §3 is thus just to nutshell certain developments that help bring about the eventual necessary/sufficient conditions for transfinite math.[1]

c. 500–c. 1200 CE Nothing much going on in Western math thanks to Rome, Aristotle, Neoplatonism, Church, etc. The real action now is in Asia and the Islamic world. By at latest 900 CE, Indian math has introduced zero as the 'tenth numeral' and the familiar goose-egg as its symbol,[2] has developed a decimal system of positional notation that is basically our base-10 own, and has codified the essentials of how 0 works in arithmetic ($0 + x = x$, $\frac{0}{x} = 0$, no fair dividing by 0, etc.). Indian and Arab mathematicians, free of any Greek geomephilia, are able to work with numbers qua numbers and to achieve significant results with negative integers, the

[1] The only other general aim of §3 is for a level of reduction and simplification that is consistently < grotesque.

[2] If you learned in school that the symbol came from the Greek omicron, you got lied to.

aforementioned 0, irrational roots, and variables to stand for arbitrary numbers and so state general properties.[3] Most of the Indo-Arab innovations make it back to Europe later, thanks mainly to Islamic conquests (e.g. of India (whose math the Arabs assimilated) in the 600s, as far west as Spain by 711, etc.).

c. 1260 St. Thomas of Aquinas's *quia* arguments for God's existence[4] constitute the official merger of Aristotelian metaphysics and Church doctrine. Thomas's basic move is to argue that since everything in the world has a cause, and those causes in turn have causes, and so on, there must at some point in the chain be an original uncaused Cause, namely God. Note, FYI, that this is essentially the same as Aristotle's famous Unmoved First Mover argument from Book VIII of the *Metaphysics*, which U.F.M. argument both Augustine and Maimonedes adopt for God-proofs, too. More important, see that for Thomas's argument to work, you have to accept the unspoken premise that an infinite transitive chain of causes and effects is impossible or incoherent. In other words, you have to regard as axiomatic the impossibility of ∞ as an actual feature of time or the universe, which basically means you're buying Aristotle's relegation of ∞ to the same weird potential-only status as Rodin's *Thinker* in all blobs of bronze.

[3] This last you will recognize as a definition of algebra, the word for which is actually a corruption of the *Al-jabra*, a treatise by the Baghdad mathematician al-Khowarizmi (d. < 850 CE).

[4] **IYI** in *Summa Theologiae* and *De Potentia Dei* (w/, **IYI**$_2$, '*quia*' meaning reasoning from effects back to causes).

Elsewhere in *Summa Theologiae*, though, Thomas advances a more original argument:

> The existence of an actually infinite multitude is impossible. For any set of things one considers must be a specific set. And sets of things are specified by the number of things in them. Now, no number is infinite, for number results from counting through a set in units. So no set of things can actually be inherently unlimited, nor can it happen to be unlimited.

This passage gets quoted by G. Cantor himself in his "*Mitteilungen zur Lehre vom Transfiniten*,"[5] wherein he calls it history's only really significant objection to the existence of an actual ∞. For our purposes, there are two significant things about Thomas's argument: (1) It treats of ∞ in terms of "sets of things," which is what Cantor and R. Dedekind will do 600 years hence (plus Thomas's third sentence is pretty much exactly the way Cantor will define a set's *cardinal number*). (2) Even more important, it reduces all of Aristotle's metaphysical distinctions and complications to the issue of whether infinite *numbers* exist. It's easy to see that what Cantor really likes here is feature (2), which makes the argument a kind of tailormade challenge, since the only really plausible

[5] = "Contributions to the Study of the Transfinite" (1887), which is one of Cantor's most important papers and appears on pp. 378–440 of his *Gesammelte Abhandlungen mathematischen und philosophischen Inhalts* (= *Collected Works*). We're citing this in such horrific detail so that from now on it will be understandable if there aren't detailed polysyllabic citations and translations-of-citations of every last Cantor-related snippet. Pretty much everything Cantorian is findable in the *Gesammelte Abhandlungen*, w/r/t which you can refer to the Bibliography for complete specs.

rebuttal to Thomas will consist in someone giving a rigorous, coherent theory of infinite numbers and their properties.

c. 1350 + Brief Time-Jump Three medium-important figures w/r/t continuity and infinite series: N. Oresme, R. ('The Calculator') Suiseth, and Fr. G. Grandi. Oresme in the 1350s invents a 'latitudinal' method for graphing motion and uniform acceleration.[6] Among other things, it affords the first hint that relative velocity (= sloped line) and relative area (= area under sloped line) are two aspects of the same thing. Around the same time, Suiseth solves a particular latitudinal problem that amounts to proving that the infinite series $\frac{1}{2}+\frac{2}{4}+\frac{3}{8}+\frac{4}{16}+\cdots+\frac{n}{2^n}$ has a finite sum, namely 2. (N.B.: No one thinks yet of applying this method to the Dichotomy.) Oresme then responds by proving that another infinite series—$\frac{1}{2}+\frac{1}{3}+\frac{1}{4}+\frac{1}{5}+\frac{1}{6}+\frac{1}{7}+\cdots+\frac{1}{n}$, alias the Harmonic Series—doesn't yield a finite sum even though the sequence of individual terms clearly appears to be approaching 0. (**IYI** Oresme's proof is ingeniously simple. By grouping the series' terms such that the first term = the first group, the second and third terms = the second group, the fourth–seventh terms = the third group, etc., so that the *n*th group contains 2^{n-1} terms, he proves that you end up with an infinite number of groups each of whose partial sums is $\geq \frac{1}{2}$, producing an infinite sum for the series.)

[6] **IYI** = a pre-Cartesian, pre-Newtonian, 100% geometrical kind of primitive calculus, with the method's eponymous latitudes being little vertical line-segments whose lengths represent velocity at an instant (the instants were the graph's 'longitudes'). Not sure whether all that explains the technique's name or just makes it more confusing. . . .

Suiseth and Oresme's series are, of course, respective examples of convergence and divergence, but no one for centuries will know how to name or handle different kinds of infinite series.[7] Even in the postcalculus era, when series became the obvious way to represent complicated functions for differentiation and integration, certain Zeno-avatars kept coming up with paradoxes that confounded various attempts to systematize convergence and divergence. One of the most fiendish of these was the good old oscillating $1 - 1 + 1 - 1 + 1 - 1 + 1 \ldots$ series from §1d, which the Catholic mathematician G. Grandi liked to use to torment the Bernoulli Brothers, famous colleagues of Leibniz who had proved the divergence of Oresme's Harmonic Series in the 1690s. Recall that the trick of the Grandi Series is that depending how you group the terms it ends up equaling both 0 and 1 . . .; or, by plugging $x = 1$ into the protologarithmic $\frac{1}{1+x} = 1 - x + x^2 - x^3 + \cdots$, you get the equality $\frac{1}{2} = 1 - 1 + 1 - 1 + 1 - 1 + 1 \ldots$, which Grandi waggishly suggested was how God had created something ($\frac{1}{2}$) from the Void (0).

(**IYI** If you've retained some high-school algebra, it's also worth looking at a nasty little divergent series that L. Euler (1707–1783, icon of early analysis) got snookered by in the 1730s. You'll recall that by the long-division rules for polynomials, $\frac{1}{1-x} = 1 + x + x^2 + x^3 + \cdots$, which if you then plug in $x = 2$ becomes the unhappy $-1 = 1 + 2 + 4 + 8 + \cdots$. Or you can also get the Grandi Series again by using $x = -1$ in the above expansion.

[7] Guess why.

Or, if your college math is solid, you can amuse yourself expanding $\frac{1}{1+x}$ by the Binomial Theorem to get $\log(1+x) = x - \frac{1}{2}x^2 + \frac{1}{3}x^3 + \cdots$ and seeing (as I. Newton, N. Mercator, and J. Wallis all did) that when $x = 2$ the series has an infinite sum but should also equal log 3. There's pretty much no end to these sorts of crunchers.)

c. 1425–35 Florentine architect F. Brunelleschi invents the technique of linear perspective in painting; L. B. Alberti's *Della pictura* is the first published account of how it works. We all probably know how paintings before the Renaissance look flat and dead and weirdly disproportionate. Brunelleschi applies geometry to pictorial space by figuring out a way to represent a 3D horizontal 'ground-plane' in a 2D vertical 'picture-plane'. The technique is most easily seen in the representations of horizontal squares (say, the floor-tiling in Florence's Baptistery) as parallelograms (in numerous pictures of same) that get flatter and sharper-angled as the floor stretches away into the painting's background. Brunelleschi/Alberti effectively conceive of a painting as a clear window interposed between a scene and the viewer, and they observe that any and all 'orthogonals,' or parallel lines receding into space at 90° to that window, will appear to converge to a vanishing point at viewer-eye-level. This vanishing point is conceived, geometrically, as a point infinitely distant from the viewer. Just about everybody knows what Masaccio, Dürer, Da Vinci, et al. were able to do with this discovery.

Mathematically, the point-at-∞ concept is later employed by J. Kepler in the Principle of Continuity that he establishes for conic sections and then uses for his Laws of Planetary Motion (see just below); and it's also central to G. Desargues's

1640-ish invention of projective geometry, and thus later to topology, to Riemannian geometry, to tensor analysis (without which in turn there'd be no General Relativity), etc. etc.

1593 The *Varia Responsa* of F. Viète (French lawyer/cryptographer) includes the first formula for summing an infinite geometric series,[8] one that's awfully close to the prenominate $\frac{a}{1-r}$ of freshman math. Although it isn't pretty, Viète is also the first to give a precise numerical expression for π, viz. as an infinite product expressible as

$$\frac{2}{\pi}=\sqrt{\frac{1}{2}}\times\sqrt{\frac{1}{2}+\frac{1}{2}\sqrt{\frac{1}{2}}}\times\sqrt{\frac{1}{2}+\frac{1}{2}\sqrt{\frac{1}{2}+\frac{1}{2}\sqrt{\frac{1}{2}}}}\times\cdots$$

(**IYI** The object of salients like these last two is to establish that ∞ in various instantiations and contexts is coming to have more and more fruitful applications even as it remains metaphysically suspect and nobody has any idea how to deal with it mathematically.)

1637 R. Descartes's *La géometrie* introduces the now-ubiquitous Cartesian coordinate plane, which lets geometrical figures be represented arithmetically/algebraically.

c. 1585–1638 Three important figures, of whom two are extremely famous: S. Stevin, J. Kepler, and G. Galilei.

In the 1580s, Stevin (Flemish engineer) resurrects Eudoxus's Exhaustion Property in deriving formulas for the weight-bearing characteristics of different geometric figures. E.g., in his *Statics* (1586) Stevin proves that a triangle's center of gravity lies on its median by inscribing an unlimited number

[8] **IYI** Viète didn't use the words 'sum,' 'geometric,' or 'series,' but in effect that's what they were.

of arbitrarily small parallelograms in the triangle and proving stuff about the centers of gravity of the resulting inscribed figures. Stevin, who is a.k.a. the Dutch Archimedes, deserves more fame than he got. Here's a decontextualized but apposite quotation from Carl Boyer: "It was largely the resulting modifications of the ancient infinitesimal methods that ultimately led to the calculus, and Stevin was one of the first to suggest these changes."

As promulgated in his *Astronomia nova* (1609), J. Kepler's 2nd Law of Planetary Motion depends on conceiving the area circumscribed by a radial vector linking an orbiting planet to the sun as composed (meaning the area's composed) of infinitely many infinitely skinny triangles, each with vertex A at the sun and vertices B & C infinitesimally close together along the orbital path. Kepler's summing the areas of this infinity of infinitesimals was, 70 years before Leibniz, applied calculus.[9]

1636–38: Galileo Galilei, under Inquisitory house arrest in Florence, produces *Two New Sciences*, a Plato-style dialogue on mechanics/dynamics. There's a whole slew of ∞-related stuff in this book. Just one example is the way Galileo applies Oresme's latitudinal graphing techniques to projectile motion and proves that the curve described by a projectile's path is a parabola. After 2000 years' mathematical study of conic sections, Kepler's orbital ellipse and G. G.'s projectile parabola

[9] **IYI** Sorry about the occluded math-prose. If you can stand it, c.f. also Kepler's 1615 *Stereometria doliorum* (= *Measurement of the Volume of Barrels* (long story, involves Emperor Rudolph II and the Austrian wine industry)), which book's 'volumetric' method for determining the areas/volumes of figures created by rotating curves entails treating solids as composed of *n* infinitesimal polygons whose areas can be summed—again, well before Newton and Leibniz.

are the first real applications of conics in physical science. Kepler's lesser-known *Ad Vitellionem paralipomena*[10] had already shown that ellipses, hyperbolas, parabolas, and circles are all products of a weird harmonic dance between two foci; the parabola is explained as what happens to a hyperbola when one focus's position relative to the other reaches ∞. Not at all accidentally, Kepler's whole theory of conics' interrelations is known as the Principle of Continuity.

Galileo's *Two New Sciences* was in certain respects one long raspberry at the Inquisition, whose treatment of G. G. is infamous. Part of this agenda was to have the dialogue's straight man act as a spokesman for Aristotelian metaphysics and Church credenda and to have his more enlightened partner slap him around intellectually. One of the main targets is Aristotle's ontological division of ∞ into actual and potential, which the Church has basically morphed into the doctrine that only God is Actually Infinite and nothing else in His creation can be. Example: Galileo ridicules the idea that the number of parts that any line segment can be divided into is only 'potentially' (meaning unreal-ly) infinite by showing that if you bend the segment into a circle—which, à la Nicholas of Cusa,[11] is defined as a regular polygon with a ∞ of sides—you have "reduced to actuality that infinite number of parts into which you claimed, while it was straight, were contained in it only potentially."

Galileo's spokesman also spends a lot of time on infinitesimals, mainly because of their utility in Stevin and Kepler's results. G. G. is the first to distinguish between different 'orders' of infinitesimals, mainly via an involved argument

[10] **IYI** It's true: all early-modern math titles sound like horrible diseases.

[11] **IYI** c. 1401–1464, mathematician and R. C. Cardinal; long story.

about why, if the earth spins, objects aren't thrown off the world at various tangents to the spin's curve, which is all a long story but the upshot is that two infinitesimals are of different orders if their ratio tends either to 0 or to ∞ and are of the same order if their ratio's finite. This is relevant because: (1) the idea that higher-order infinitesimals are so unbelievably tiny and evanescent that they can be discarded from an equation because they'll have no effect on the result ends up being vital for classical calculus; and (2) Galileo's distinction anticipates some of G. Cantor's own discoveries about the strange arithmetic of infinite quantities, viz. that not all ∞s are the same size but that the differences between them aren't really arithmetical (e.g., adding n to ∞ doesn't increase it, nor does adding ∞ to ∞ or multiplying ∞ by ∞) but more like geometric.[12]

The extreme mathematical weirdness of ∞, which Galileo spends a lot of time in *TNS* giving examples of, is rather presciently attributed to epistemology instead of metaphysics. Paradoxes arise, according to G. G.'s mouthpiece, only "when we attempt, with our finite minds, to discuss the infinite, assigning to it those properties which we give to the finite and limited." The big illustration of this is §1's Paradox of Galileo, in which recall you can set up a one-to-one correspondence between all the integers and all the perfect squares even though it's evident that there are way more integers than perfect squares.[13] From this cruncher, Galileo concludes that "we must say that there are as many squares as there are

[12] **IYI** This all gets unpacked in much more detail in the climactic §7 below.

[13] **IYI** In fact, the ratio $\frac{\text{Total \# of integers}}{\text{Total \# of P.S.s}}$ itself tends to ∞ as you get farther and farther out in the sequence.

numbers," and thus (again) that "the attributes 'equal,' 'greater,' and 'less' are not applicable to infinite, but only to finite quantities." Though the latter conclusion turns out to be wrong, *TNS*'s is still the first truly modern attitude toward actual infinities as mathematical entities. Notice, for instance, that Galileo does not pull out the old Aristotelian reductio and conclude from the paradoxical behavior of infinite sets that ∞ can't be reasoned about. Instead, he manages somewhat to anticipate both Kant (by attributing ∞-paradoxes to the hardwired constraints of 'finite minds' rather than to any extramental reality) and Cantor (by using one-to-one correspondence as a comparative measure of sets, by arguing that infinite quantities obey a different sort of arithmetic than do finite quantities, etc.).

Familiar fact: The seventeenth century, with its Counter-Counter-Reformation and Scientific Revolution, saw the first real explosion in philo-mathematical progress since the Hellenistic acme. This is the century in which Descartes invents coordinate geometry (as well as Radical Doubt), Desargues invents projective geometry, Locke empiricism, Newton and Leibniz college math. None of these would have been possible without a loosening of the Aristotelian stranglehold on Western thought. Galileo's *TNS* is up there with Descartes's *Discourse on Method* and Bacon's *Novum Organum* in terms of holdbreaking, and it is no accident at all that so much of its time is spent on ∞. From among a whole ream of apposite supporting quotations here, see Prof. T. Danzig's "When, after a thousand-year stupor, European thought shook off the effect of the sleeping powders so skillfully administered by the Christian Fathers, the problem of infinity was one of the first to be revived."

The other way *Two New Sciences* is important is in its sustained and original use of the *function*. You doubtless remember what a mathematical function is and why it's hard to define clearly (such as, e.g., 'A relation between variables,' 'A rule for establishing the image of a domain,' 'A mapping'). A function is at least one abstraction-level up from variables, being basically a rule for pairing elements in one set with elements in another set. For now, let's assume we all pretty much know what a function is—or rather what it *does*, since a function is really a kind of procedure even though symbolism like '$f(x) = \frac{1}{x}$' tends to make it seem like a thing. At least graphically, the idea of a function had been around since Oresme in the fourteenth century, although Oresme had used Scholastic terminology and called his technique a *latitude of forms*, 'form' being the Aristotelian term for features or qualities, which were thought to include things like the speed of a moving body. Not until Galileo would people understand that velocity is not a quality of the thing moving, but rather an abstract process representable by the schoolboy function $r = \frac{d}{t}$, just as (up one level of abstraction) it was G. G. who determined that acceleration operates as the function $s = \frac{1}{2}at^2$.

Two New Sciences is the first math book to make extensive nongraphical use of functions, although the functions here are described verbally and often (à la the Greeks) in terms of proportions and ratios. What's striking is the speed with which the concept/theory of functions gained currency once a certain critical mass of needs and permissions was reached.[14] Most of these needs involved continuity. The broad outline

[14] J. Gregory (q.v. main text just below) gives the first widely accepted definition of a function in a book on quadrature problems only 30 years after *TNS*.

here is that Kepler's astronomy and Galileo's studies of motion—which were themselves motivated largely by the need for improved timekeeping in navigation (again, long story)—created the impetus for a rigorous study of curves, which curves the Cartesian coordinate plane allowed to be expressed algebraically, i.e. as functions like $y = x^2$, $y = \sin x$, and so on. The important distinctions between polynomial, algebraic, and transcendental functions[15] were easily derived from Descartes's classifications of curves, as well as from the explicit representations of functions by different kinds of

[15] The differences between these sorts of functions can be left at the fact that the transcendentals are the really hairy ones: trigonometric, exponential, logarithmic, etc. What can't be left vague is the synonymic distinction between algebraic and transcendental *numbers*, which is part of the whole broad taxonomy in which of course integers + fractions compose the rational numbers, rational + irrational numbers make up the real numbers, a real number plus an imaginary number like $\sqrt{-1}$ constitutes a complex number, and so on. Given our general purposes, we don't have to deal with anything beyond real numbers, luckily. But be advised that the reals' irrational component itself comprises two different kind of numbers—or rather the distinction rational v. irrational kind of overlaps another distinction, that between algebraic and transcendental numbers. This difference becomes important when we get to Cantor's proofs about the various sizes of the ∞s of different 'number-classes'. So: An *algebraic number* is one that is the root of a polynomial with integer coefficients. As in, say, $\sqrt{8}$ is an algebraic number because it's the zero-root of $1x^2 - 8 = 0$. (Actually integers, rationals, and even complex numbers can be algebraic too—like e.g. the respective zero-roots of $2x - 14 = 0$, $2x - 7 = 0$, and $3x^2 - 2x + 1 = 0$—but in terms of Cantor/Dedekind/continuity we need nail down only the surds.) *Transcendental numbers*, then, are those that aren't algebraic, i.e. that can't be roots of integer-coefficient polynomials; π is a transcendental surd, as is *e*, the base of natural/hyperbolic logarithms (which don't worry about it if that term's not familiar).

series a little later (= c. 1670). The word 'function' itself, by the way, comes from G. W. Leibniz.[16] Which of course is hardly an accident, since Leibniz helped invent the calculus, and one of calc's most powerful features is the use of functions to represent processes. After Leibniz, the crevasse-fraught concept of 'continuous phenomena' gets replaced in math by the continuous function and the infinite series . . . and in point of fact G. Cantor's explorations of ∞ will end up coming out of a particular application of just these tools to a certain set of problems involving heat. Which is obviously a very long story we're now engaged in trying to set up.

Here, by the way, is a quotation from D. Berlinski: "It is the contrast between the continuous and the discrete that is the great generating engine by which the real numbers are constructed and the calculus created." Just so we remember where we are in the overall forest; this section is just trees.

c. 1647–65 Three medium-important figures, who if this had been 200 years earlier would all now be extremely famous: Gregory of St. Vincent, J. Wallis, and J. Gregory.

c. 1647: Gregory of St. V. proposes a solution to Zeno's Dichotomy that explicitly mentions the sum of a geometric series.[17] He's also the first mathematician to posit that an infinite series represents an actual magnitude or sum, which he is also the first to posit as the series' limit, which he calls the "progression's terminus" and describes in rather Eudoxian

[16] **IYI** 'Function' was Leibniz's alternative to Newton's weird word 'fluent'; and, as was the case with a lot of other terminology, Leibniz's term became the preferred one. Factoid: Leibniz also introduced 'constant' and 'variable'.

[17] **IYI** The proffered solution is actually to "Achilles v. the Tortoise," but it amounts to the same thing.

terms as an end "to which the progression does not attain, even if continued to infinity, but to which it can approach more closely than by any given interval."

1655: Wallis, the 2nd-greatest British mathematician of the century, publishes his aforementioned *Arithmetica infinitorum*, whose title is 0% coincidental. This is the first major work on infinite series' application to the arithmetization of geometry, and it will be indispensable to Newton's version of calculus[18] a couple decades hence. Among *A.i.*'s important results: The first correct general def. of the limit of an infinite sequence and the sum of an infinite series; the use of an infinite product to represent the sine and cosine; the demonstration that $\frac{\pi}{2} = (\frac{2 \times 2}{1 \times 3}) \times (\frac{4 \times 4}{3 \times 5}) \times (\frac{6 \times 6}{5 \times 7}) \times \cdots$ (compare Leibniz's $\frac{\pi}{4} = 1 - \frac{1}{3} + \frac{1}{5} - \frac{1}{7} + \frac{1}{9} - \frac{1}{11} \cdots$ a few years later); and of course the first use of '∞' as a symbol for ∞.

1665: J. Gregory (a Scot) defines 'function,' lobbies to make *approaching a limit* the sixth basic function of algebra, and expands several different trig and inverse trig functions into infinite series, e.g. proving that '$\arctan x = x - \frac{x^3}{3} + \frac{x^5}{5} - \frac{x^7}{7} + \cdots$' holds for $-1 \leq x \leq 1$. A whole lot of series-expansion work goes on around this time, mostly because navigators, engineers, et al. needed much more detailed and accurate trigonometry- and logarithm tables, and expansions of functions into infinite series was the best way to interpolate table values.

[18] **IYI** This is because Anglican calc depended so heavily on infinite series and the Binomial Theorem—q.v. §4a below.

(**IYI Also c. 1665** The Binomial Theorem (i.e., the high-school formula for expanding $(p + q)^n$) is liberated from the $(p + q)^{n-1}$-dependence of Pascal's Triangle by I. Newton. The expansion is thought to be infinite for fractional or negative *n*'s, but nobody can really prove anything about the B.T. or the convergence/divergence of series in general until J.-B. J. Fourier in the 1820s.)

§3b. As has been at least implied and will now be exposited on, the math-historical consensus is that the late 1600s mark the start of a modern Golden Age in which there are far more significant mathematical advances than anytime else in world history. Now things start moving really fast, and we can do little more than try to build a sort of flagstone path from early work on functions to Cantor's infinicopia.

Two large-scale changes in the world of math to note very quickly. The first involves abstraction. Pretty much all math from the Greeks to Galileo is empirically based: math concepts are straightforward abstractions from real-world experience. This is one reason why geometry (along with Aristotle) dominated mathematical reasoning for so long. The modern transition from geometric to algebraic reasoning[19] was itself a symptom of a larger shift. By 1600, entities like zero, negative integers, and irrationals are used routinely. Now start adding in the subsequent decades' introductions of complex numbers, Napierian logarithms, higher-degree polynomials and literal coefficients in algebra—plus of course eventually the 1st and 2nd derivative and the integral—and it's clear that as

[19] **IYI** This change is symbolized nicely by trigonometry's movement from degrees and geometric shapes to radians and trig functions.

of some pre-Enlightenment date math has gotten so remote from any sort of real-world observation that we and Saussure can say verily it is now, as a system of symbols, "independent of the objects designated," i.e. that math is now concerned much more with the logical relations between abstract concepts than with any particular correspondence between those concepts and physical reality. The point: It's in the seventeenth century that math becomes primarily a system of abstractions from other abstractions instead of from the world.

Which makes the second big change seem paradoxical: math's new hyperabstractness turns out to work incredibly well in real-world applications. In science, engineering, physics, etc. Take, for one obvious example, calculus, which is exponentially more abstract than any sort of 'practical' math before (like, from what real-world observation does one dream up the idea that an object's velocity and a curve's subtending area have anything to do with each other?), and yet is unprecedentedly good for representing/explaining motion and acceleration, gravity, planetary movements, heat—everything science tells us is real about the real world. Not at all for nothing does D. Berlinski call calculus "the story this world first told itself as it became the modern world." Because what the modern world's about, what it *is*, is science. And it's in the seventeenth century that the marriage of math and science is consummated, the Scientific Revolution both causing and caused by the Math Explosion because science—increasingly freed of its Aristotelian hangups with substance v. matter and potentiality v. actuality—becomes now essentially a mathematical enterprise[20] in which force,

[20] **IYI** Newton obviously the bestriding Colossus here. . . .

motion, mass, and law-as-formula compose the new template for understanding how reality works. By the late 1600s, serious math is a part of astronomy, mechanics, geography, civil engineering, city planning, stonecutting, carpentry, metallurgy, chemistry, hydraulics, hydrostatics, optics, lens-grinding, military strategy, gun- and cannon-design, winemaking, architecture, music, shipbuilding, timekeeping, calendar-reckoning: everything.

And the practical influence cuts both ways. Here is a definitive quotation from M. Kline: "As science began to rely more and more upon mathematics to produce its physical conclusions, mathematics began to rely more and more upon scientific results to justify its own procedures." And, as will be made heavy weather of in §§ 4 and 5 below, this union is fruitful but also rife with hazards. In brief, all sorts of formerly dubious quantities and procedures are now admitted to math on account of their practical efficacy, meaning that if mathematics wants to retain its deductive rigor they will have to be rigorously 'theorized' and grounded in math's axiomatic schema. Guess which examples of these long-questionable concepts we're interested in here. Have a look at the pellucid Kline again, here in a chapter of his *Mathematical Thought* titled "Mathematics as of 1700": "Infinitely large quantities, which the Greeks had studiously avoided, and infinitely small ones, which the Greeks had skillfully circumvented, [now] had to be contended with."

§3c. So then once the Story of ∞ hits the late 1600s we're now barreling at high and irreversible speed toward Cantor et al., and the math gets a lot more abstract and technical. And a Command Decision's been made that at selected points you are going to have to be subjected to quick little EMERGENCY GLOSSARIES in which certain unavoidable terms/concepts are defined so that they can then be used without constantly having to stop and noodle around *in medias* about what they mean. Some will be new; some have already been mentioned or may seem sort of obvious but are important enough that they and some of their associated subterms have to be nailed down 100% tight.

N.B.: The following first EMERGENCY GLOSSARY may be a bit dry due to sheer compression; and though it was tempting to designate it **IYI** for readers with strong math backgrounds, the fact is that many of the definitions are so radically decocted and simplified that it's probably worth your time to at least scan E.G.I so you're clear on the specific ways we're going to be using the terms. For readers without much college math, on the other hand, the following should be all we need to proceed for at least the next few §s.

EMERGENCY GLOSSARY I, W/ AN ASSOCIATED NARRATIVE TIME-JUMP

—*Real Line* As mentioned, this is essentially an amped-up Number Line, meaning a geometric line with a fixed dense scale so that every real number corresponds to a unique point on the line. For our purposes, the Real Line is a 'topological space,' which here means that the Line and the set of all real numbers it represents can be used interchangeably to refer to the same abstract thing[21]—which thing, it's also already been mentioned, is usually called 'The Continuum,' w/ this term

[21] N.B., especially for later, that Weierstrass's, Dedekind's, and Cantor's theories about real numbers and continuity are often referred to in math books as *the topology of the Real Line.*

itself meaning exactly what it looks like it means: the combined origin and instantiation of continuity.

—*Function* Pretty much already covered in §3a—or have a look at this wonderful def. straight from a 5th-grade math class: "A relationship between two things where the value of one is determined by the value of the other." You'll recall from basic algebra that in a regular function like $y = f(x)$, x is the *independent variable* and y is the *dependent variable*, meaning simply that changes in x produce other changes in y according to the rules of f. The set[22] of all possible values that can be assumed by the independent variable is called the function's *domain*; the set of all possible y-values is the function's *range.*

—*Real Function* A function whose domain and range are sets of real numbers.

—*Continuous Function* (a) The function $y = f(x)$ is continuous if tiny little changes in x yield only tiny little changes in y; there are no big jumps or gaps or weirdnesses. If a function is *discontinuous*, it's usually discontinuous *at* a certain value for the independent variable; e.g. $f(x) = \frac{x^2 - 1}{x - 1}$ is discontinuous at $x = 1$.[23] (FYI, there happen to be all different kinds of discontinuities, each with its own characteristic behavior and

[22] **IYI** sets of course being strictly speaking post-Cantor entities, but what are you going to do. . . .

[23] **IYI** That is, if you graph $f(x) = \frac{x^2 - 1}{x - 1}$, you'll see that the resultant curve has a hole in it corresponding to 1's position on the x-axis, because here the $f(x)$ equals $\frac{0}{0}$, which is mathematically defined as mathematically undefined. (**IYI**$_2$ The especially savvy reader might notice that there's limit-value and limit-of-function stuff going on in this example, too, which we're not mentioning because we haven't done *limits* yet.)

graph-shape and technical name—'jump discontinuity,' 'removable discontinuity,' 'infinite discontinuity'—but we're probably not going to fool with these distinctions.)

—*Interval* The amount of space on the Real Line between two points, say p and q, which is equivalent to the set of all real numbers between p and q. Here p and q are called the interval's *endpoints.* The *closed interval* $[p, q]$ contains the endpoints; the *open interval* (p, q) doesn't. Notice the brackets for closed intervals and the parens for open ones; that's how the difference gets symbolized.

—*Neighborhood* On the Real Line, the *neighborhood* of a point p is the open interval $(p - a, p + a)$ where $a > 0$. Another way to express this is to say that the *a-neighborhood* of p is the set of all points whose distance from p is less than a.

—*Continuous Function* (b) Functions are often ID'd as continuous/discontinuous *in* or *over* certain intervals. A function $f(x)$ is continuous over the open interval (p, q) if it is continuous at every point in (p, q). For it to be continuous over the closed interval $[p, q]$, the following have to be true:

$$\lim_{x \to p^+} f(x) = f(p) \quad \text{and} \quad \lim_{x \to q^-} f(x) = f(q)$$

which of course will make sense only if you're conversant with limits.

—*Limits* Or rather maybe *Limits* v. *Bounds,* since these are related but also crucially different. The distinction is probably easiest to see with respect to sequences. Oops.

—*Sequence* Any succession of terms formed via some rule, e.g. the geometric sequence 1, 2, 4, 8, 16, . . ., 2^{n-1}, ...

—*Limits* v. *Bounds* (a–d) The informal mnemonic that Dr. G. always suggested was that *limit* involves the expressions 'tends to' or 'approaches,' while *bound* takes the modifier 'upper' or 'lower'. (a) The *limit of a sequence* is the great unspoken concept behind Zeno's Dichotomy and is implicit in Eudoxian Exhaustion, Kepler's volumetrics, etc. W/r/t sequences, 'limit' refers to the number you never actually arrive at but do get closer and closer and closer to as the number of terms in the sequence grows. Put a little more sexily, the limit L of the infinite sequence $p_1, p_2, p_3, \ldots, p_n, \ldots$ is the number that the sequence approaches (or 'tends to') as n approaches ∞, with this latter approach symbolized by a little sublinear '$\rightarrow$' and the whole thing by $\lim\limits_{n\rightarrow\infty} p_n = L$. (b) The *limit of a function* is basically the value that the dependent variable approaches as the independent variable approaches some other value. A ubiquitous Calc I example is $f(x) = \frac{1}{x}$, where $f(x)$ approaches 0 as x approaches ∞, written as $\lim\limits_{x\rightarrow\infty}(\frac{1}{x}) = 0$.[24] (c) The *bound of a function* is a totally different horse. It's a restriction of some kind(s) on the function's range. A classic example from trig is $f(x) = \sin x$, where all the values of $f(x)$ are going to be between -1 and 1. More important for our purposes is that functions can have *upper bounds* (U) and/or *lower bounds* (L) such that $f(x) \leq$ U and/or $f(x) \geq$ L for all x in the function's domain. Even more important are the further-specified *least upper bound* and *greatest lower bound* of a function, where U_1 is the least upper bound of $f(x)$ if any other upper bound U_n is $\geq U_1$, and L_1 is $f(x)$'s greatest lower bound if it's $\geq$ any other lower bound L_n. (d) Sequences can

[24] **IYI** The 'approaches' stuff is actually technically wrong, as will get spelled out in some detail when we start talking about Weierstrassian analysis in §5e. The idea here in E.G.I is to make limits intuitively clear, not mathematically rigorous.

have *bounds* in pretty much the same way as functions. The infinite sequence of positive integers 1, 2, 3, . . . obviously has a lower bound at 0, which will also constitute the upper bound of $-1, -2, -3, \ldots$. A *bounded sequence* is one that's got both an upper and lower bound; e.g. if $x \geq 1$, it's easy to see that the sequence generated by expanding[25] $1 - (\frac{1}{x})$ will be so bounded.[26,27]

—*Series* Definable as a sequence whose terms are all added to one another, as in the geometric series $1 + 2 + 4 + 8 + 16 + \cdots + 2^{n-1} + \cdots$. The intimate relation of series to sequences means that they share most qualities and associated predicates, with one big exception: where sequences have limits, series have both limits and *sums.* You might recall the infamous Big Sigma of college math, which lets you designate the sum even of series with an infinite number of terms—because it turns out that all the interesting series are infinite. The sum of the infinite series $p_1 + p_2 + p_3 + \cdots + p_n \ldots$ is written $\sum_{n=1}^{\infty} p_n$, where the tiny antipodal '∞' and '$n = 1$' indicate the *limits* (meaning here the range of possible values of n) of the series.[28] Infinite series are *convergent* if they

[25] oops$_2$: see —*Expansion* below

[26] **IYI** viz. by a lower 0 and an upper 1, which you're welcome to start plugging in values for x and see for yourself.

[27] Please N.B. now that bounds and boundedness work pretty much the same way for sets as they do for sequences. This will start being important in §7, at which time you will probably be asked to flip back and review this very FN.

[28] **IYI** More lagan for later retrieval: If it's occurred to you to wonder whether the sum's appogiaturan '∞' denotes an actual limit or end or rather in fact the *absence* of any limit/end, be advised that this question is

converge to a finite sum (see, e.g., how the Zenoid $1 + \frac{1}{2} + \frac{1}{4} + \frac{1}{8} + \cdots$ converges to the sum 2) and *divergent* if they don't (as in the series $1 + 2 + 3 + 4 + \cdots$); but both kinds of series have at least abstract sums[29] that can be symbolized via 'Σ' and treated as quantities in further calculations. —*Infinite product* Sort of like an infinite series except the terms are multiplied.[30] A lot of things in trigonometry, from π to the sine and cosine functions, can be represented as infinite products, depending on how you handle the expansions. —*Expansion* This means putting something mathematical in the form of a sequence/series/product (we're particularly interested in series-expansions). How it works depends on

---

highly significant and goes to the heart of what Weierstrass/Dedekind/Cantor were able to do for analysis (w/r/t which stay tuned for —*Analysis*). If, on the other hand, you're wondering how mathematicians before Weierstrass actually viewed these ∞s that their *x*'s and *n*'s 'approached,' the basic answer is that they relegated these infinitely large/small quantities to the same vague, dotted-outline existence as Aristotle's potential infinite. The idea is that the real mathematical/metaphysical status of limits' infinities never has to be considered, because nothing ever actually gets there. If this strikes you as a bit shifty, then you are already in a position to see why Weierstrass thought it all needed rigorizing.

[29] Shit. All right. The strict truth is more complicated than that, and involves the limits of sequences of partial sums, where a *partial sum* = the sum of some finite number of consecutive terms in a series. The basic idea is that if the infinite sequence of its partial sums tends to some limit *S*, then an infinite series is convergent and its sum is *S*. And that a divergent series is one whose sequence of partial sums doesn't approach a limit, and so it doesn't have a finite sum. All of which is way too abstract at this point but hopefully will make more sense by the end of §5.

[30] **IYI** We're not going to be worrying about associated terms like *continued product* and *oscillating product*.

what you're expanding. The expansion of a mathematical expression is usually pretty straightforward, as in recall all the mechanical $(x + y)^2 \rightarrow x^2 + 2xy + y^2$ operations of high-school math (w/r/t which you'll recollect also that whatever constants there are in front of the terms' variables are known as the series' *coefficients*). Functions, on the other hand, are more interesting and thus more complicated. Not all are even expandable, for one thing. For a function to be *representable* as a series, the function's series-expansion either (1) must be finite or (2) must, if it's infinite, converge to the function for all values of the variables. Example: The trig function cos x is representable by the convergent power series $1 - \frac{x^2}{2!} + \frac{x^4}{4!} - \frac{x^6}{6!} + \frac{x^8}{8!} - \ldots$[31]

—*Power Series* A particular kind of series involving exponents (a.k.a. powers), the generic power-series form being $p_0 + p_1x + p_2x^2 + p_3x^3 + \ldots + p_nx^n + \ldots$ where the x-values are real numbers and the p-values are coefficients. Factoid: The expansions of the basic sine, cosine, elliptic, hyperbolic, log, and exponential functions are all power series (as too is Zeno's Dichotomy).

—*Fourier Series,* which are sort of the sum of two power series, are 3rd- or 4th-term college math[32] and can be real brainmelters, but they're vital to the context of transfinite math and have to be at least generally pinned down. For our purposes, Fourier Series can be regarded as expansions of *periodic functions,* w/r/t which latter all you need to know is that they're ways to represent various kinds of waves and so are also sometimes called *wave functions.* The fundamental wave functions are trig's sin x and cos x, and the elementary

[31] **IYI** where the factorial '2!' means 2×1, '4!' means $4 \times 3 \times 2 \times 1$, etc.

[32] **IYI** usually studied under the heading Harmonic Analysis.

Fourier Series is the expansion of a trig function $f(x)$ into—get ready—$\sum_{n=0}^{\infty} a_n \cos(nx) + \sum_{n=0}^{\infty} b_n \sin(nx)$,[33] where a and b are what's known as *Fourier coefficients*, which are so conceptually hairy that we plan to avoid them at almost any cost.

—*Quadrature* This is the 1600s' term for a certain kind of problem that led to integral calculus. Technically, it refers to constructing a square whose area = the area bounded by a closed curve. An early-modern version of the old Squaring the Circle problem, in other words. We're bothering to define 'quadrature' because it gets used below in certain historical contexts where it would be wrong to say 'integration' instead because integration did not, strictly speaking, exist yet.

—*Derivative* (*n.*) The McGuffin of differential calc. In sexual terms, it's an expression of the rate of change of a function with respect to the function's independent variable.[34] Since it might ring bells from math class, let's add that the derivative of a function $f(x)$ at a certain point p can be understood as the slope of the tangent to the curve given by $y = f(x)$ at p, although this won't have much application for us. Important bonus factoid: The process of finding a given derivative is called *differentiation.*

[33] **IYI** If by chance you are flipping back to this entry from §5b and noticing that this F.S. looks different from Fourier's original Exhibit, be apprised that the two are really the same. It's just that the above form makes it clearer how Fourier Series comprise and combine two different trig series, which—if you're not flipping back—will in turn make more sense when *trigonometric series* are defined in E.G.II. Sorry if this is confusing; we're doing the best we can.

[34] **IYI** If you're innocent of college math, this def. will make more sense when we look at classical calc in §4a.

—*Integral* (*n.*) This is the inverse of the derivative, i.e. the function that has a given derivative, i.e. the function the derivative's derived from; i.e., if $f(z)$ is the derivative of $g(x)$, then $g(x)$ is the integral of $f(x)$. Way more on all this in an actual context coming up in §4. (N.B. The process of finding a given integral(s) is called *integration*, which is what mathematicians often do when they're stuck on a problem and don't know how to proceed. Hence the calligraphic slogan in many math-grad-student offices: *DON'T SIT AND WAIT*—***INTEGRATE.***)

—*Analysis* Another highly abstract term that can't be finessed or avoided. There's a very formal definition involving the way certain types of functions vary around the neighborhood of a point on a surface, which given our overall agenda can be dispensed with in favor of the idea that analysis is the branch of math that studies anything having to do with limits or 'limiting processes'—meaning calculus, functions of real and complex variables, topology of the R.L., infinite sequences and series, and so on. Books and classes often refer to analysis as 'the mathematics of continuity.' Which can be a little misleading, because most of us are also taught that continuity is the jurisdiction of calculus, and there are some wholly non-calc areas that are still analysis, of which areas a couple are especially relevant. Algebra sort of bleeds into analysis via the Binomial Theorem[35] when n is < 0 and the expansion of $(p + q)^n$ becomes the infamous Binomial Series; likewise trig → analysis when, e.g., the sine- and cosine functions are expanded into their respective power series.

An additional complication is that for modern mathematicians 'analysis' can also connote a particular sort of

[35] **IYI** which got de facto defined at the very end of §3a.

methodological spirit in which the above sorts of fields are studied. See e.g. this from the *Oxford Concise Dictionary of Mathematics*: "The term 'analysis' has come to be used to indicate a rather more rigorous approach to the topics of calculus, and to the foundations of the real number system," in which the latter parallel phrase = the purview of Dedekind and Cantor, w/ the reasons why calc topics should have needed "a more rigorous approach" constituting the real motive cause behind their work. In brief, the whole rigor-and-foundations thing was part of the great philosophical emergency of postcalculus math, a deep split over how mathematical entities should be viewed and theorems proved; and this split is in turn the deep context behind the controversies over Cantor's transfinite math. All of which will get hashed out as we proceed.

Something else implicit in the *Oxford* quotation involves the old oppositions discrete v. continuous and geometry v. pure math. As it happens, the big names behind the early calculus were all concerned with continuous functions and magnitudes that were either outright geometric (lines, curves, areas, volumes) or geometrically representable (force, velocity, acceleration). Be advised now, though, that one of math's major preoccupations in the century leading up to Cantor and Dedekind will be the *Arithmetization of Analysis*, which in essence means deriving theorems about continuous functions using only numbers, not curves or areas. This Arithmetization ends up bringing analysis more into the provinces of algebra and number theory, fields that had hitherto been devoted to 100% discrete math entities/phenomena. What occurs in nineteenth-century analysis will be a detachment from geometry similar to Greek math's after the D.B.P.'s discovery of irrationals.

We are now once again sort of out over our skis, chronologically speaking. The main ∞-related question to keep in mind throughout the next couple §s is going to be why exactly calculus should have required the additional rigor mentioned above in the EMERGENCY GLOSSARY (which we're now more or less no longer in). It's also worth emphasizing, again for future use, that the most important distinction between discrete and continuous phenomena in math is that the former can be characterized with just rational numbers, whereas continuity requires all the reals, meaning also irrationals.

It so happens that an important figure in both the Arithmeticization of Analysis and the mathematics of ∞ is Fr. B. P. Bolzano (1781–1848) of the University of Prague, whom for a variety of reasons this is the place to talk about—although to do this we're going to have to get briefly into the 1800s and then kind of hiccup and go backwards again in the next §. Arithmeticization-wise, the priest Bolzano is the least well-known of a quartet of mathematicians who pioneered what came to be known as 'rigorous analysis' in the early nineteenth century, the other three being A.-L. Cauchy, N. Abel, and P. G. L. Dirichlet. Cauchy tends to get the most credit, thanks mainly to his *Cours d'analyse* (1821), which became the standard college-math textbook in Europe for 150 years. Broadly stated, Cauchy's project involves trying to rescue calculus from its metaphysical difficulties[36] by defining infinitesimals rigorously in terms of limits; but much of Cauchy's analysis is still beholden to geometry in ways that end up causing problems. It's actually Bolzano, in his 1817 *Rein analytischer Beweis des Lehrsatzes . . .*,[37] who gives the first purely arithmetical proof of a theorem involving continuous

[36] **IYI** re which, again, see directly below and §4 even more below.

[37] **IYI** = "Purely Analytical Proof. . . ." The full title is 22 words long and you do not want it.

functions.[38] In this same book he supplies what is now considered to be the correct mathematical definition of continuity: $f(x)$ is continuous in interval A if at any point a in A the difference $f(a + \delta) - f(a)$ can be made as small as you want by making δ arbitrarily small. What Bolzano really is is another vivid instance of math-fame's caprice. Some of this will be contextlessly *ante rem* here, but be apprised that, for example, his method for determining whether a series is continuous still gets used today—and is attributed to Cauchy. Or that Bolzano was the first mathematician to come up with a function that's continuous but not differentiable (i.e., has no derivative), a result that overturned early calc's assumption that continuity and differentiation went hand in hand—and was completely ignored, w/ K. Weierstrass's construction of a similar function 30 years later getting hailed as its 'discovery'.[39]

All this will end up being more important than it looks right now, particularly the idea of continuity as an arithmetical property. It's Bolzano's later work on infinite quantities[40]

[38] The specific proof is that algebraic polynomials are continuous, which is less relevant than the connections between a function being continuous in an interval and a series/sequence of functions being convergent in an interval. These connections start becoming really important in §5.

[39] To get still further ahead of ourselves: Discussed at some length in §5e will be an important hypothesis of Bolzano's about infinite sequences and *limit points*, a hypothesis that Weierstrass also rediscovered and proved, though here history has thrown B. B. a bone and called this the Bolzano-Weierstrass Theorem, which Theorem as it happens is important to R. Dedekind's theory of irrational numbers. (FYI, there's no suggestion that Weierstrass ripped Bolzano off or anything. These sorts of parallel discoveries happen all the time in math.)

[40] namely his 1851 *Paradoxien des Unendlichen*, which wasn't even available in English (as *Paradoxes of the Infinite*) until 1950.

that's apposite here, though, if only because it's the most important historical link between Galileo's *Two New Sciences* and the work of Dedekind/Cantor. For one thing, Bolzano (who was kind of a heretic, both mathematically and religiously (e.g., he eventually got dismissed from U. Prague for giving anti-war sermons)) is the first mathematician since Galileo to address explicitly the distinction between Aristotle's actual and potential infinities. Like *TNS*, Bolzano's *Paradoxes of the Infinite* is deeply anti-Aristotelian, though there are also important differences—Bolzano's arguments are a lot more mathematical than G. G.'s, as is the arguments' motive. To mention once again some stuff that's going to get developed in more detail below, the impetus behind *P. of the I.* has to do with certain metaphysical difficulties involved in calculus's deployment of ∞-related quantities and increments. Pretty much all postcalculus mathematicians had tried to dodge or obfuscate these difficulties by vaguely invoking Aristotle and assuming that all the ∞s they were tossing around were only potential or 'incomplete' (this was the basic idea behind Cauchian limits). That Bolzano attempted to blow large ragged holes in this assumption is one reason why his work got so little attention. It's also why Prof. G. Cantor, who tends to be uniformly venomous about most historical treatments of ∞, often singles Bolzano out for special praise.[41]

P. of the I. is a product of Bolzano's combined interests in functions, infinite collections, and the Real Line. As it happens, the book is only a few concepts short of inventing

[41] **IYI** See for example the English version of Cantor's "Foundations of the Theory of Manifolds," which is in the Bibliography under ARTICLES & ESSAYS.

modern set theory—'modern' in the sense of being able to handle infinite sets.[42] One of the big ways it anticipates Cantor's work is that it renders overt something that was only implicit in the Paradox of Galileo, namely the idea of one-to-one correspondence as a way to establish the equivalence of two sets. Bolzano's approach to Galileo's Paradox is purely abstract, and Cantorian. It consists in taking something that Galileo had attributed to limitations of the human mind and making it an intrinsic property of infinite sets—viz. the fact that a subset of an infinite set can have as many members as the set itself. As we're going to see, after G. Cantor (whose own work was controversial but not at all neglected), mathematicians understood that this property was in fact the distinctive feature of infinite sets; and the formal math definition of *infinite set* is now based on this freaky equivalence.

Notice also that Galileo's version of the equivalence concerned only the Big ∞s of all integers and all perfect squares. It's Bolzano who first formulates the equivalences between the dense, Zenoish Little ∞s of the Real Line. He does this in *P. of the I.* by examining the set of all real numbers between 0 and 1, i.e. the set of all points in the closed interval [0,1] on the R.L. Bolzano sets up the elementary function[43] $y = 2x$ and observes that if its domain's values of x are all the points in [0, 1], the function will assign to each x one and only one y-value in the larger closed interval [0, 2]. So that .26 will correspond to .52, .74 to 1.48, .624134021 . . . to 1.248268042 . . .,

[42] As we'll see in §7, most of formal set theory is trivial if you assume that only finite sets exist.

[43] Like Cantor, Bolzano has a gift for giving simple, pictorially compelling proofs of very abstract propositions.

and so on. In other words, a perfect one-to-one correspondence: There are exactly as many R.L.-points in [0, 1] as there are in [0, 2]. And (as now appears obvious, but Bolzano was the first to point it out) simply by changing the function's x-coefficient to any other integer—$y = 5x$, $y = 6{,}517x$—you can prove that there are exactly as many real numbers between 0 and 1 as between 0 and any other finite number you can think of.*

***IYI-GRADE INTERPOLATION**

Actually, as was tossed off in §2e, the number of points in [0, 1] is ultimately equal to the ∞ of points on the whole Real Line stretching infinitely in both directions. Though the formal proof of this is pretty involved,[44] a demonstration of the equivalence is within the capacities of the average 4th-grader. Take the Real Line–segment corresponding to [0, 1] and stick it above the entire R.L., then place a compass's pointy part on the segment's exact midpoint and draw the lower half of a circle C whose diameter is 1,[45] and arrange it all like so:

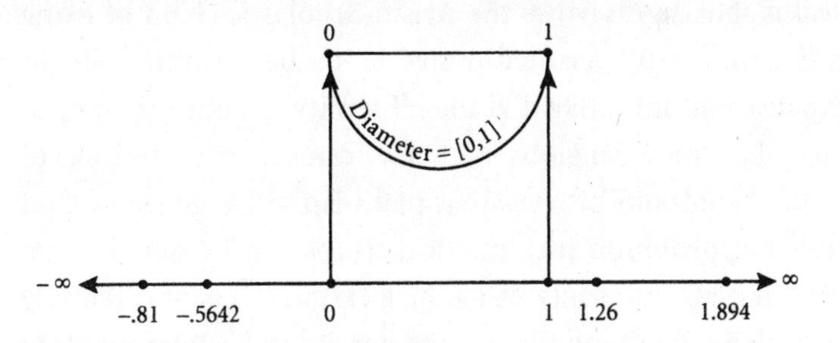

[44] **IYI** q.v. §7d; or, if you're flipping back from §7d, now you can see why we went into this equivalence in such detail here.

[45] meaning the diameter's the [0, 1] interval itself. (**IYI** The length of the hemispheric arc will obviously be $\frac{\pi}{2}$, but we don't really care about that.)

Pick any point on the Real Line and draw a straight line L from that point to the center of C, i.e. to the diameter's midpoint. Wherever L hits the hemisphere, draw a line straight up to hit the [0, 1] diameter, again like so:

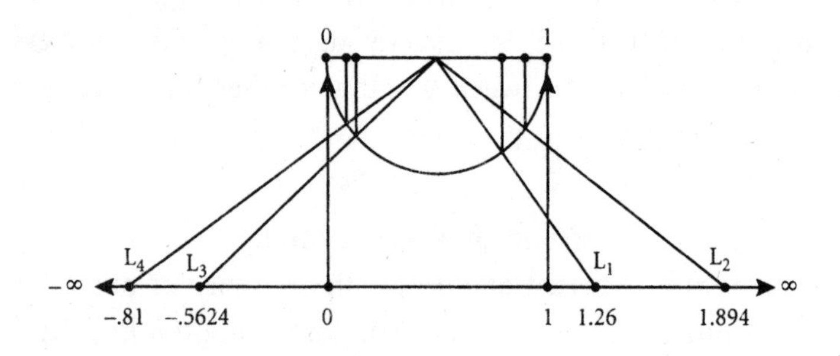

Thus every point on the R.L. can, via an L_1, L_2, L_3, . . ., L_n, be shown to correspond one-to-one with a point in [0,1]. Q.E.D.

END INTERPOLATION

Technical prescience aside, *Paradoxes of the Infinite* is also remarkable for its metaphysical agenda. In this, too, it resembles *TNS* and some of Cantor's later work. Bolzano's basic deal is that he disavows the Aristo-Scholastic chain of being and believes the created universe to be both infinite in expanse and infinitely divisible. 'Eternity' is simply a temporal ∞. Like most religious mathematicians from Pythagoras to Gödel,[46] Bolzano believes that math is the Language of God and that profound metaphysical truths can be derived and proved mathematically. What he lacks, in terms of extending his insights about infinite size and density and equivalence into actual theorems, are the set-theoretic concepts of *cardinality,*

[46] This includes Cantor in some of his less guarded and/or stable pronouncements.

ordinality, and *power* as they apply to collections of points.[47] He can establish and prove the strange equivalence of infinite sets/subsets, and can foresee that their relation is not contradictory but paradigmatic; but he has no way to turn his proofs into an actual theory of infinite sets and their relations, behavior, etc. The main reason for this—strange though it may sound right now—is that in Bolzano's era there isn't yet any coherent theory of the real-number system, no rigorous def. of *irrational number*.

§4a. The scholarly consensus is that there have been three big periods of crisis in the foundations of Western math. The first was the Pythagorean incommensurables. The third is the era (which we're arguably still in) following Gödel's Incompleteness proofs and the breakdown of Cantorian set theory.[1] The second great crisis surrounded the development of calculus.

The idea now is going to be to trace out how transfinite math gradually evolves out of certain techniques and problems associated with calculus/analysis. In other words, to build a kind of conceptual scaffold for viewing and appreciating G. Cantor's achievements.[2] As mentioned, this means

[47] **IYI** Again, these all get defined and discussed in §7.

[1] Command Decision: We're going to quit saying 'see below' all the time and simply assume that from here on it will be obvious when it applies.

[2] **IYI** As mentioned, the rhetorical aim here is to rig the discussion so that it's not grotesquely reductive but is simple and clear enough to be followable even if you've had no college math. It's true that it would be *nice* if you've had some college math, but please rest assured that considerable

going diachronically back to where the original timeline left off at the end of §3a.

So then now we're near the end of the seventeenth century, the time of the Restoration and the Siege of Vienna, of towering perukes and scented hankies, etc. Doubtless you know already that calculus was the most important mathematical discovery since Euclid. A seminal advance in math's ability to represent continuity and change and real-world processes. Some of this has already been talked about. You probably also know that I. Newton or/and G. W. Leibniz are usually credited with its discovery.[3] You might also know—or at least have been able to anticipate from §3a's timeline—that the idea of exclusive or even dual credit is absurd, as is the notion that what's now called the calculus comprises any one invention. By even the simplest accounting, royalties would need to be shared by a good dozen mathematicians in England, France, Italy, and Germany who were all busily ramifying Kepler and Galileo's work on functions, infinite series, and the properties of curves, motivated (meaning the mathematicians were, as has also been mentioned) by certain pressing scientific problems that were also either math problems or treatable as same.

Here were some of the most urgent motivating problems: calculating instantaneous velocity and acceleration (physics,

---

pains have been taken and infelicities permitted in order to make sure it's not required.

[3] In fact there was a big row in European math over which one of them really invented it, specifically over whether Leibniz, whose first calc-related paper was in 1674, had plagiarized Newton, whose *De analysi per aequationes numero terminorum infinitas* circulated privately in 1669.

dynamics); finding the tangent to a curve (optics, astronomy); finding the length of a curve, the area bounded by a curve, the volume bounded by a surface (astronomy, engineering); finding the maximum/minimum value of a function (military science, esp. artillery). There were probably some other ones, too. We now know that these problems are all closely related: they're all aspects of calculus. But the mathematicians working on them in the 1600s didn't know this, and Newton and Leibniz do deserve enormous credit for seeing and conceptualizing the relations between, for example, the instantaneous velocity of a point and the area bounded by its motion's curve, or the rate of change of a function and the area given by a function whose rate of change we know. It was N. & L. who first saw the forest—meaning the Fundamental Theorem that differentiation and integration are mutually inverse—and were able to derive a general method that worked on all the above-type problems. On the mystery of continuity itself. Although not without having to dance around some nasty crevasses in this forest, and certainly not without all sorts of other people's preliminary arboreal results and discoveries. Those in addition to the ones already timelined include, e.g.: 1629—P. de Fermat's method for finding the max. and min. values of a polynomial curve; 1635ish—G. P. de Roberval's discovery that a curve's tangent could be expressed as a function of the velocity of a moving point whose path composed the curve[4]; 1635—B. Cavalieri's Method of Indivisibles for calculating the areas under curves; 1664—I. Barrow's geometrical Method of Tangents.

[4] Sorry about the hideous syntax here; there's no nice way to compress Roberval.

Plus, c. 1668, there's a great prescient ¶ in the preface to J. Gregory's *Geometriae Pars Universalis* whose upshot is that the really important division of math is not into the geometrical v. the arithmetical but into the *universal* v. the *particular.* Why this is prescient: Various mathematicians from Eudoxus to Fermat had invented and deployed calc-type methods, but always geometrically and always in relation to specific problems. It's Newton and Leibniz who combine the various methods of Latitudes and Indivisibles & c. into a single arithmetic technique whose breadth and generality (i.e., its abstractness) are its great strength.[5] The two's backgrounds and approaches are different, though. Newton comes to calculus via Barrow's Method of Tangents, the Binomial Theorem, and Wallis's work on infinite series. Leibniz's route involves functions, patterns of numbers called 'sum-' and 'difference-sequences,' and a distinctive metaphysics[6] whereby a curve could be treated as an ordered sequence of points separated by a literally infinitesimal distance. (In brief, curves for Leibniz are generated by

[5] This is why trying to settle the credit question by saying that Newton invented differential calculus and Leibniz invented integral calculus (which some math teachers like to do) is confusing and wrong. The whole point is that N. & L. understood the Problem of Tangents (= instantaneous velocity) and the Problem of Quadratures (= areas under curves) to be two aspects of a single larger problem (= that of continuity) and thus treatable by the same general method. The whole reason N. & L. are math immortals is that they didn't split calc up the way intro courses do.

[6] Leibniz, like Descartes, being also of course a big-time philosopher, of whose ontology you may have heard terms like 'individual substance,' 'transcreation,' 'identity of indiscernibles,' and 'windowless monad'.

equations, whereas quantities varying along a curve are given by functions (pretty sure we've mentioned that he copyrighted 'function').)

We're not going to get too much into the Newton-v.-Leibniz thing, but the metaphysical differences in the way they viewed infinitesimal quantities are highly germane.[7] Newton, at heart a physicist who thought in terms of velocity and rate of change, used infinitely tiny increments in his variables' values as disposable tools in arriving at the derivative of a function. Newton's derivative was basically a Eudoxian-type limit of these increments' ratio as they got arbitrarily small. Leibniz, a lawyer/diplomat/courtier/philosopher for whom math was sort of an offshoot hobby,[8] had an aforementionedly idiosyncratic metaphysics that involved certain weird, fundamental, infinitely small constituents of all reality,[9] and he pretty much built his calculus around the relations between them. These differences had methodological implications, obviously, with Newton seeing everything in terms of rates of change and the Binomial Theorem and thus tending to represent functions[10] as infinite series, v. Leibniz preferring what are known as 'closed forms' and avoiding series in favor of summations and straight functions, including transcendental functions when algebraics wouldn't work. Some of these differences were just taste—e.g., the two used

[7] **IYI** Some of the following might be a bit eyeglazing in the abstract, but it will make more sense shortly when we look at a simple example.

[8] Surely we all hate people like this.

[9] **IYI** these being the monads mentioned three FNs up.

[10] **IYI** even functions involved in area problems.

totally different notations and vocab, although Leibniz's was better and mostly won out.[11] For us, the important thing is that both men's versions of calculus caused serious problems for mathematics as a deductive, logically rigorous discipline, and were vigorously attacked at the same time that they enabled all sorts of incredible results in math and science. The source of the foundational shakiness should be easy to see, whether the problem appears more methodological (as in Newton's case) or metaphysical (as in G. W. L.'s). As has been mentioned in §2b and probably elsewhere (and is well known anyway), the trouble concerns infinitesimals, which all over again in the late 1600s force everybody to try to deal with the math of ∞.

The best way to talk about these problems is to sketch the way early calculus works. We're going to do a somewhat non-standard, quadrature-type derivation that manages to illustrate several different aspects of the technique at once so that you don't have to sit through a bunch of different cases. We're also going to sort of mix and match N. & L.'s different methods and terminology, since the aim here isn't historical accuracy but clarity of illustration. For the same reason, we'll eschew the usual how-to-find-the-tangent or how-to-go-from-average-speed-to-instantaneous-speed cases most textbooks use.[12]

Refer first to what we'll call Exhibit 4a, which please note isn't even remotely to scale but does have the advantage of

[11] **IYI** Among other Leibnizisms are 'differential calculus,' 'integral calculus,' 'd*x*,' and the good old vermiculate integral sign '∫,' which latter (Gorisian factoid:) Leibniz originally meant as an enlarged S denoting "the sum of the [*y*-coordinates] under a curve."

[12] N.B. re possible **IYI**: If you've got a strong math background, feel free to skip the following Exhibit and gloss altogether—the simplifications may bother you more than it's worth.

making stuff easy to see. For the same reasons, the relevant 'curve' in Exhibit 4a is a straight line, the very simplest kind of curve, w/r/t which the calculations are minimally hairy.[13] E4a's curve here can be regarded either as a set of points produced by a continuous function on a closed interval or as the path of a moving point in 2D space. For the latter, Newtonian case (which is what most college classes seem to prefer), note that here the vertical axis indicates position and the horizontal axis is time, i.e. that they're reversed from the axes in the motion-type graphs you're apt to have had in school (long story; good reasons). So:

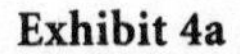
Exhibit 4a

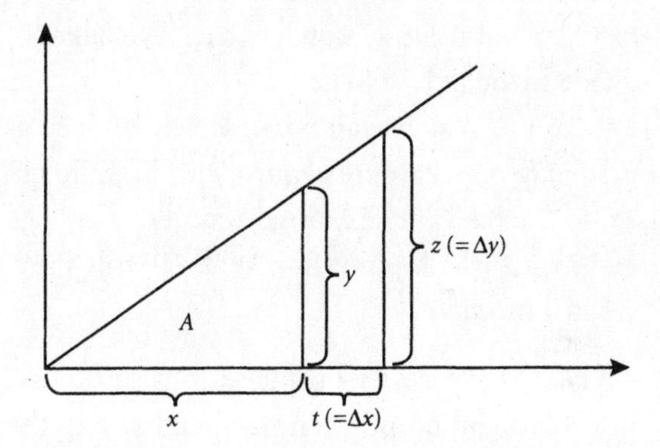

First, posit that A, the area under the curve, is equal to x^2. (This will seem strange because E4a's curve is a straight line,

[13] **IYI** For readers w/ strong backgrounds who nevertheless haven't skipped all this but are noticing already that Exhibit 4a looks like a very simplified illustration of Leibniz's "difference quotient," and are maybe wondering why we don't just go ahead and do his famous Characteristic Triangle, the answer is that using the C.T. would cause the problems' explication to eat 6+ pages and subject everybody to too much calc-detail that ends up not being important.

so it looks like A really ought to be $\frac{xy}{2}$; but for most curves drawn exactly to scale x^2 is going to work, so here please just play along and pretend that y exactly equals x.) Meaning formally we assume that:

(1) $A = x^2$. Then posit that x increases by some infinitesimally tiny quantity t,[14] with the area under the curve consequently increasing by tz. Given this, and given the equality in (1), we have:

(2) $A + tz = (x + t)^2$. Multiplying out (2)'s binomial, we get:

(3) $A + tz = x^2 + 2xt + t^2$. And since, by (1), $A = x^2$, we can reduce (3) to:

(4) $tz = 2xt + t^2$. Now watch close. We take (4) and divide through by t to get:

(5) $z = 2x + t$. Watch again: since we've defined t as *infinitesimally tiny*, $2x + t$ is equivalent, in finite terms, to $2x$, so the relevant equation becomes:

(6) $z = 2x$. At which point fini. What-all this shows we'll see in a moment.

You will likely have noticed some serious shiftiness in this derivation's treatment of infinitesimal quantity t. In the move from (4) to (5), t is sufficiently > 0 to be a legal divisor. In the move from (5) to (6), though, t appears to be $= 0$, since t added to $2x$ yields $2x$. In other words, t is being treated as 0 when it's convenient and as > 0 when it's

[14] Again, if E4a is treated as a rate-of-change problem, t is an infinitesimal instant. (**IYI** If you've encountered the somewhat unfashionable term *infinitesimal calculus* in connection with classical calc, you can now see that the term derives from the infinite tininess of quantities/durations like t.)

convenient,[15] which appears to create the contradiction $(t = 0)$ & $(t \neq 0)$, which—as you'll recall from the previous discussion of reductio-type proofs—seems like ample grounds for going back and saying there's got to be something wrong with using infinitesimal quantities like *t*. At the very least, the *t* thing looks like a notational trick, some math version of Cooking the Books in accountancy.[16]

Except here's the thing. If you blink the apparent contradiction, or at least hold off on running a reductio on it, a derivation like E4a's (which, notwithstanding its resemblance to Leibniz's Characteristic Triangle, is actually a simplified version of the process Newton uses in *De Analysi*[17]) turns out to be a truly marvelous piece of mathematical ordnance, one

[15] N.B. that according to Leibniz this is precisely what infinitesimals are—they're critters you can do this with. See for example this excerpt from a letter to J. Wallis around 1690:

> It is useful to consider quantities infinitely small such that when their ratio is sought, they may not be considered zero but which are rejected as often as they occur with quantities incomparably greater. Thus if we have $x + [t]$, $[t]$ is rejected. But it is different if we seek the difference between $x + [t]$ and x. . ..

[16] An even better analogy might be an experimental scientist Skewing his Data to confirm whatever hypothesis he wants confirmed.

[17] **IYI** Newton's examples in *D.A.* were messier and depended more on the Binomial Theorem, whereby an equivalence like $z = rx^n$ (where r is a constant and the n may be a fraction or even negative) can be expanded to show that rx^n's rate of change will always $= nrx^{n-1}$. This is what allows for the theoretically infinite chain of *higher derivatives* in college math. As in the 1st derivative of, e.g., $y = x^4$ is $4x^3$; the 2nd derivative is $12x^2$, and so on, until any *n*th derivative can be found via the ratio $\frac{d^n y}{d^n x}$—although you usually don't get into anything higher than 2nd derivatives in regular calc.

that yields at least two crucial results. Result #1 is that the rate of change of x^2 can be shown to be $2x$ if you accept the computation $\frac{(x+t)^2 - x^2}{t}$ as representing the change in x during the 'instant' t.[18] Result #2 is that you can show the rate of change of area A to be the 'curve' (namely E4a's y (remember that a straight line is a kind of curve)) that bounds A. To see this, compute $\frac{A + tz - A}{t}$ and cancel to get $\frac{tz}{t}$, then divide through by the suspiciously convenient t to get z, which remember is only 'infinitesimally greater' than y and so here can be regarded as $= y$.[19] You end up with $y = 2x$, which happens to be the function that produces Exhibit 4a's curve. Which means that packed into the result $y = 2x$ is the basic principle of integral calc: the rate of change of the area bounded by a curve is nothing other than that very curve. Which in turn means that the integral of a function that has a given derivative is the function itself, which happens to be the Fundamental Theorem of the Calculus,[20] viz. that differentiation and integration are inversely related[21] the same way

[18] Observe, though, that you have to use the same sort of questionable accounting practices in this calculation:

(1) $\frac{(x+t)^2 - x^2}{t} =$

(2) $\frac{x^2 + 2tx + t^2 - x^2}{t}$, which $=$

(3) $\frac{2tx + t^2}{t}$, at which point you assume that $t \neq 0$ and divide through to get:

(4) $2x + t$, whereupon you assume that $t = 0$ and toss it out to get:

(5) $2x$.

[19] Or you can get the same result by treating z as effectively equivalent to y in equation (6) of the original derivation.

[20] IYI as first articulated by Leibniz in 1686.

[21] IYI This is why syllabiphiles call integration 'antidifferentiation'.

multiplication/division and exponents/roots are, which is why the calculus is so powerful and N. & L. deserve so much credit—the F.T.C. combines both techniques in one high-caliber package (. . . so long as you accept the equivocations about whether $t = 0$).

This is, however, not the way most of us had these matters explained in school. If you took Calc I, chances are that you learned, via velocity-and-acceleration graphs, to 'take the limit of Δx,' or that '$\frac{dy}{dx} = \lim_{x \to \infty} \frac{\Delta y}{\Delta x}$,' with '$\frac{dy}{dx}$' being Leibniz's notation and the limit concept being postcalc analysis's subsequent way of finessing the whole problem of infinitesimals. You might, for example, know or recall that in most modern textbooks an infinitesimal is defined as 'a quantity that yields 0 after the application of a limit process'. If you are an actual Calc I survivor, you surely can also remember how brutally abstract and counterintuitive the limits thing is to try to learn: almost nobody ever tells undergraduates the whys or whences of the method,[22] or mentions that there's an easier or at least more intuitive way to understand dx and Δx and

[22] In this context, what the limits method really is is a metaphysical accounting trick that makes infinitude/infinitesimality a feature of the calculation process rather than of the quantities calculated. As should be evident by now, the regular laws of arithmetic don't work on ∞-related quantities; but by basically restricting itself to partial sums through 99% of the calculation, limits-based calculus lets these rules apply. Then, once the basic calculation is completed, you 'take the limit' and let t or dx or whatever 'approach 0,' and extrapolate your result. In pedagogical terms, the math student is asked here to presume that certain quantities are finite and stable for calculation purposes but then vanishingly tiny and protean at the actual results stage. This is an intellectual contortion that makes calculus seem not just hard but bizarrely and pointlessly hard, which is one reason why Calc I is such a dreaded class.

t, namely as orders of $\frac{1}{\infty}$. Most teachers instead try to distract students with snazzy examples of calculus's ability to solve all kinds of complex real-world problems—from instantaneous velocity and -acceleration as 1st and 2nd derivatives, to Kepler's elliptical orbits and Newton's F = m(dx), to the motions of sprung springs and bouncing balls, eclipses' penumbrae, loudness as a function of a volume-knob's rotation; not to mention the trigonometric vistas that open up when you learn that d(sin x) = cos x and d(cos x) = − sin x, that the tangent is the limit of the secant, etc. These are usually presented as the inducements for mastering the limits concept, a concept that is really no less abstract or algesic than trying to conceive of dx or t as just incredibly, mindbendingly tiny.

As should be understandable from the foregoing, the true motive behind the limits approach to calc was that Newton and Leibniz's infinitesimal quantities and notational sleights of hand had opened up some nasty cracks in math's foundations, given that the proposition '($x = 0$) & ($x \neq 0$)' violates all sorts of basic LEM-ish axioms. Given our Story so far, the easiest thing to say appears to be that most of the supposed problems here were actually caused by math's inability to handle infinite quantities—that, as with Zeno's Dichotomy and Galileo's Paradox, the real difficulty was that no one yet understood the arithmetic of ∞. It wouldn't exactly be wrong to say this, but for our purposes it would be at least semi-impoverished.[23] As with everything else about math after calculus, the real problems and stakes here are more complex.

[23] **IYI (and maybe not even a good idea to mention)** There is, as a matter of fact, a nontrivial way to say the same thing, but it involves *non-standard analysis*, which is the invention of one A. Robinson in the '70s

§4b. Let's restate and summarize a bit. The sheer power and significance of calculus presented early-modern math with the same sort of crisis that Zeno's Dichotomy had caused for the Greeks. Except in a way it was worse. Zeno's Paradoxes hadn't solved any extant problems in math, whereas the tools of calculus did. The panoply of real-world results that calc enabled has already been detailed, as has the extraordinary timing—every kind of applied science is butting up hard against problems of continuous phenomena just as Newton and Leibniz and their respective cadres come up with a mathematical account of continuity.[24] One that works. One that leads directly to the great modern decoding of physical laws as differential equations.

Except foundationally it's a disaster. The whole thing's built on air. The Leibnizians[25] couldn't explain or derive actual quantities that were somehow not 0 but still infinitely close to 0.

---

and professes to rigorize infinitesimals in analysis via the use of *hyperreal numbers*, which themselves basically combine the real numbers and Cantorian transfinites—meaning the whole thing's heavily set-theoretic and Cantor-dependent, plus controversial, and wildly technical, and well beyond this discussion's limits . . . but nongrotesqueness appears to require at least mentioning it, and maybe commending w/r/t any burning further interest on your part Prof. Abraham Robinson's *Nonstandard Analysis*, Princeton U. Press, 1996.

[24] This probably needs to be explained instead of just asserted over and over. In classical calculus, continuity is treated as essentially a property of functions: a function is continuous at some point *p* if and only if it's differentiable at *p*. This is why Bolzano and Weierstrass's finding those continuous but nondifferentiable functions in the 1800s will be such a big deal, and why modern analysis's theory of continuity is now a lot more complicated.

[25] **IYI** meaning mainly the two J. Bernoullis and J_1's son D., plus G. F. A. l'Hôpital, who'd been one of the J.s' patron (the Bernoullis are hard to keep straight), all flourishing in, say, the early 1700s.

The Newtonians,[26] who claimed that calculus didn't really depend on infinitesimal quantities but rather on 'fluxions'—w/ *fluxion* meaning the rate of change of a time-dependent variable—foundered on the requirement that the ratios of these fluxions be taken just as they vanish into or emerge from 0, meaning really the infinitesimal first or last instant when they're > 0, which of course is just trading infinitely tiny quantities for infinitely brief instants. And the Newtonians had no better account of these instantaneous ratios than the Leibnizians did of infinitesimal quantities.[27] The only real advantage

[26] **IYI** = primarily the U.K.'s E. Halley, B. Taylor, and C. Maclaurin, also early 1700s.

[27] **IYI** It so happens that Bishop G. Berkeley (1685–1753, major empiricist philosopher and Christian apologist (and a world-class pleonast)) has a famous critique of classical calc along just these lines in an eighteenth-century tract whose 64- (yes, 64-) word title starts with "*The Analyst*. . . ." A representative snippet being:

> Nothing is easier than to devise expressions or notations for fluxions and infinitesimals. . . . But if we remove the veil and look underneath, if, laying aside the expressions, we set ourselves attentively to consider the things themselves which are supposed to be expressed or marked thereby, we shall discover much emptiness, darkness, and confusion; nay, if I mistake not, direct impossibilities and contradictions.

Berkeley's broadside is in some ways Christianity's return-raspberry to Galileo and modern science (and it's actually great cranky fun to read, though that's neither here nor there). Its overall point is that eighteenth-century math, despite its deductive pretensions, really rests on faith no less than religion does, i.e. that "[H]e who can digest a second or third fluxion, a second or third [derivative], need not, methinks, be squeamish about any point in divinity."

On the other hand, M. J. l.R. d'Alembert (1717–1783, big post-calc mathematician and all-around intellectual, plus one of the first proponents of the idea that "the true metaphysics of the calculus is to be found in the idea of a limit") objects to infinitesimals on wholly logical LEM grounds in the famous *Encyclopédie* he coedited with D. Diderot in the 1760s, as in e.g.: "A

of the Newtonian version (for everyone but Calc I students) is that it already has the limit concept sort of implicitly contained in the idea of a vanishingly tiny first/last instant—it would be mostly A.-L. Cauchy, then later K. Weierstrass, who drew all this out. (Weierstrass, by the way, was a teacher of Cantor's.)

Apropos continuity and infinitesimals and calculus, it's worth looking quickly at one more of Zeno's anti-motion Paradoxes. This one's usually called the Arrow, because it concerns the time-interval during which an arrow is traveling from its bow to the target.[28] Zeno observes that at any specific instant in this interval, the arrow occupies "a space equal to itself," which he says is the same as its being "at rest." The point is that the arrow cannot really be moving at an instant, because motion requires an interval of time, and an instant here is not an interval; it's the tiniest temporal unit imaginable, and it has no duration, just as a geometric point has no dimension. And if, at each and every instant, the arrow is at rest, then the arrow does not ever move. In fact nothing whatsoever moves, really, since at any given instant everything is at rest.

There's at least one implicit premise in Zeno's argument, which schematizing helps make overt:

(1) At each and every instant, the arrow is at rest.
(2) Any interval of time is composed of instants.
(3) Therefore, during any interval of time, the arrow isn't moving.

---

quantity is something or nothing; if it is something, it has not yet vanished; if it is nothing, it has literally vanished. The supposition that there is an intermediate state between these two is a chimera."

[28] **IYI** The Arrow, like the Dichotomy, gets discussed in Book VI of Aristotle's *Physics*; it also appears in fragmentary form in Diogenes Laërtius's *Lives and Opinions*.

The covert premise is (2), which is just what Aristotle attacks in the *Physics*, dismissing the whole Z.P. on the grounds that "... time is not composed of indivisible instants,"[29] i.e. that the very notion of something being either in motion or at rest at an instant is incoherent. Notice, though, that it is precisely this idea of motion at an instant that N. & L.'s calculus is able to make mathematical sense of—and not just general motion but precise velocity at an instant, not to mention rate-of-change-in-velocity at an instant (= acceleration, 2nd derivative), rate-of-change-in-acceleration at an instant (= 3rd derivative), etc.

Arrow-wise, the fact that classical calc is able to handle precisely what Aristotle says can't be handled is not a coincidence. First off, have another glance at the thing about an instant having "no duration" two ¶s back, and see that this term is somewhat ambiguous. It turns out that the kind of instant Zeno is talking about is, at least mathematically, not something of 0 duration, but an infinitesimal. It has to be. Consider again the hoary old jr.-high formula for motion, Rate × Time = Distance, or $r = \frac{d}{t}$. An arrow at rest has an r of 0 and covers 0d, obviously. But if, timewise, an instant = 0, then Zeno's scenario ends up positing 0 as the divisor in $r = \frac{d}{t}$, which is mathematically illegal/fallacious in the same way the whole '$0 = \frac{0}{0}$' is illegal/fallacious.

[29] **IYI** It goes without saying that the compatibility of this claim with Aristotle's time-series objections to the Dichotomy in §2b is somewhat dubious. There are ways to reconcile A.'s two arguments, but they're very complicated, and < 100% convincing—and anyway that's hardcore Aristotle-scholar stuff and well outside our purview.

Except here we have to be careful again, just as with the other Z.P.s. As has been mentioned in all kinds of different contexts already, there's 'handling' something v. really *handling* it. Even if we grant that Zeno's instant is an infinitesimal and thus ripe for treatment by Newtonian fluxion or Leibnizian d*x*, you can probably already see that a classical-calc-type 'solution' to Zeno's Arrow is apt to be trivial in the same way that '$\frac{a}{(1-r)}$' is trivial w/r/t the Dichotomy. That is, the Arrow is really a metaphysical paradox, and it's precisely a metaphysical account of infinitesimals that calc hasn't got. Without such an account, all we can do is apply to the Arrow some sexy-looking formula that will depend on the same mysterious and paradoxical-looking infinitesimals that Zeno's using in the first place; plus there will still be the unsettling question of how an arrow actually *gets* to the target over an interval comprising infinitely many $\frac{1}{\infty}$-size instants.[30]

The problem is that where the Arrow is metaphysical it is also extremely subtle and abstract. Consider for instance another hidden premise, or maybe a kind of subpremise that's implicit in Zeno's (1): is it really true that something's got to be either moving or at rest? At first it certainly looks true, provided we take 'at rest' to be a synonym for 'not moving'. Remember LEM, after all. Surely, at any given instant t, something is either moving or else not moving, meaning that it has at t either a Rate > 0 or a Rate $= 0$. That in truth this disjunction is *not* valid—that LEM doesn't really apply here—can be seen by examining the difference between the number

[30] **IYI** Feel free to review §2a's harangue about applying formulas v. truly solving problems, which applies here in spades.

0 and the abstract word 'nothing'. It's a tricky difference, but an important one. The Greeks' inability to see it was probably what kept them from being able to use 0 in their math, which cost them dearly. But 0 v. nothing is one of those abstract distinctions that's almost impossible to talk about directly; you more have to do it with examples. Imagine there's a certain math class, and in this class there's a fiendishly difficult 100-point midterm, and imagine that neither you nor I get even one point out of 100 on this exam. Except there's a difference: you are not in the class and didn't even take the exam, whereas I am, and did. The fact that you received 0 points on the exam is thus irrelevant—your 0 means N/A, nothing—whereas my 0 is an actual zero. Or if you don't like that one, imagine that you and I are respectively female and male, both healthy and 20–40 years of age, and we're both at the doctor's, and neither of us has had a menstrual period in the past ten weeks, in which case my total number of periods is nothing, whereas yours here is 0—and significant. End examples.

So it's simply not true that something's always got to be either 0 or not-0; it might instead be nothing, N/A.[31] In which case there's a nontrivial response to Zeno's premise (1), to wit: the fact that the arrow is not moving at *t* does not mean that its *r* at *t* is 0 but rather that its *r* at *t* is nothing. That this slipperiness in premise (1) is not spotted right away is due in part to the 0-v.-nothing thing and in part to the vertiginous, Level-Four abstractness of words like 'movement' and 'motion'. The noun 'motion,' for example, is especially sneaky because it doesn't look all that abstract; it seems straightforwardly to denote some single thing or process—whereas, if you think

[31] **IYI** Regarding d'Alembert's objection to infinitesimals in FN 27 supra, it is just this third possibility that makes his argument unsound.

about it,[32] even the simplest kind of motion is really a complicated relation between (a) one object, (b) more than one place, and (c) *more than one instant.* Upshot: The fallacy of the Arrow lies in Zeno's assumption that the question 'Is the arrow in motion or not at instant *t*?' is any more coherent than 'What was your grade in this class you didn't take?' or 'Is a geometric point curved or straight?' The right answer to all three is: N/A.[33]

Granted, this response to the Arrow is, strictly speaking, philosophical rather than mathematical. Just as a classical-calc-type solution will be philosophical, too, in the sense of having to make metaphysical claims about infinitesimals. Modern analysis's own way of dealing with this Z.P. is very different, and purely technical. If, again, you ever did the Arrow in college math, you probably learned that Zeno's specious premise is (1) but heard nothing[34] about an instant as an infinitesimal. This (again) is because analysis has figured out ways to dodge both the infinitesimal and the 0-as-divisor problem in its representations of continuity. Hence, in a modern math class, premise (1) is declared false because the arrow's *r*-at-instant-*t* can be calculated as 'the limit of average *r*'s over a sequence of nested intervals converging to 0 and always containing *t*,' or something close to that. Be apprised that

[32] **IYI** Here's a nice example of where some horizontal early-morning abstract thinking can really pay off. Once we're up and about and using our words, it's almost impossible to think about what they really mean.

[33] **IYI** Observe, please, that this is not at all the same as Aristotle's objection to 'Is the arrow . . . instant *t*?' What he thinks is really incoherent is premise (2)'s idea that time can be composed of infinitesimal instants, which is an argument about temporal continuity, under which interpretation the Arrow can be solved with a simple calc formula. As you've probably begun to see, Aristotle manages to be sort of grandly and breathtakingly wrong, always and everywhere, when it comes to ∞.

[34] (not 0)

the language of this solution[35] is Weierstrassian: it's his refined limit-concept[36] that will allow calculus to handle the related problems of infinitesimals and Zeno-type infinite divisibility.

The specific relations between these problems are intricate and abstract, but for us they're totally apropos. However weird or foundationally corrosive infinitesimals are, it turns out that their disqualification from math/metaphysics creates some wicked little crevasses as well. Example: Without infinitesimals, it apparently makes no sense to talk about the 'next instant' or 'very next split-second'—no two instants can be quite successive. Explanation: Without infinitesimals, then respecting any two supposedly successive instants t_1 and t_2, there are only two options: either there's no (meaning 0) temporal interval between t_1 and t_2, or there's some temporal interval > 0 between them. If there's 0 interval, then t_1 and t_2 are clearly not successive, because then they're the exact same instant. But if there is some temporal interval between them, then there are always other, tinier instants between t_1 and t_2—because any finite temporal interval can always be subdivided tinier and tinier, just like distances on the Number Line.[37] Meaning

[35] which solution, though 100% technical, at least has the advantage of recognizing that motion-at-an-instant is a concept that always involves more than one instant.

[36] **IYI** As we'll see in §5, what Weierstrass basically does is figure out how to define limits in a way that eliminates the 'tends to' or 'gradually approaches' stuff. Expressions like these had proved susceptible to Zenoid confusions about space and time (as in 'approaches from where?' 'how fast?' etc.), besides being just generally murky.

[37] This is the rub, and why the relation between infinitesimals and Zeno-type divisibility is sort of like that between chemo and cancer. The thing about quantities that are less than $\frac{1}{2}, \frac{1}{4}, \frac{1}{8}, \frac{1}{16}, \frac{1}{32}, \ldots, \frac{1}{n}$ but still greater than 0 is that you cannot get to them by dividing over and over and over again—

there's never going to be a *very next* t_2 after t_1. In fact, so long as infinitesimals are non grata, there must always be an infinite number of instants between t_1 and t_2. This is because if there were only a finite number of these intermediate instances, then one of them, t_x, would by definition be the smallest, which would mean that t_x was the instant closest to t_1, i.e. that t_x was the very next instant after t_1, which we've already seen is impossible (because of course what about the instant $\frac{t_x - t_1}{2}$?).

If you're now noticing a certain family resemblance among this no-successive-instant problem, Zeno's Paradoxes, and some of the Real Line crunchers described in §2c and -e, be advised that this is not a coincidence. They are all facets of the great continuity conundrum for mathematics, which is that ∞-related entities can apparently be neither handled nor eliminated. Nowhere is this more evident than with $\frac{1}{\infty}$s. They're riddled with paradox and can't be defined, but if you banish them from math you end up having to posit an infinite density to any interval,[38] in which the idea of succession makes no sense and no ordering of points in the interval can ever be complete, since between any two points there will be not just some other points but a whole infinity of them.

Overall point: However good calculus is at quantifying motion and change, it can do nothing to solve the real paradoxes of continuity. Not without a coherent theory of ∞, anyway.

the same way you can't get to a transfinite number by adding or multiplying finite numbers. ∞ and $\frac{1}{\infty}$ are uniquely exempt from all the paradoxes of infinite subdivision and expansion . . . even though they are in a sense the very embodiments of those paradoxes. So the whole thing is just very strange.

[38] meaning interval in time, in space, or on the Real Line—all three are continuity's turf.

§5a. The job of these next couple sections is to show how Dedekind's and Cantor's innovations arose in more or less the same way calculus had—viz. as ways to handle certain problems which had become so pressing that math couldn't really advance without facing them. The idea in §5 is to sketch out the particular post-calc developments and controversies that create an environment in which transfinite math becomes possible, which is also to say necessary. Plus please notice the 'sketch out'. There's no way to do a timeline, even a rough one, of the period 1700–1850. Too much happens too fast.

In general, the situation of mathematics after 1700 is intensely weird, and much of the weirdness has to do once again with the relations between empirical reality and conceptual abstraction.[1] As anyone who's moved from high-school to college math can attest, analysis is exponentially more abstract and difficult than anything that comes before.[2] At the same time, its explanatory power is unprecedented and its practical applications go through the roof. This is mainly because of analysis's ability to quantify motion and

[1] If we stick in 'only more so' after every predicate, the last three ¶s of §3b can be made to apply here quite nicely all over again.

[2] **IYI** This difficulty, despite what Humanities majors often think, is not because of all the heavy-looking notation that can make flipping through a college math book so intimidating. The special notation of analysis is actually just a very, very compact way to represent information. There aren't that many different symbols, and compared to a natural language it's ridiculously easy to learn. The problem isn't the notation—it's the extreme *abstractness* and *generality* of the information represented by the symbolism that makes college math so hard. Hopefully that makes sense, because it's 100% true.

process and change, and because of the greatly increased generality of physical laws expressed as differential equations and/or trigonometric series. At the *same* same time, and just as with the development of classical calc, much of the mathematical progress from 1700 to at least 1830 is in response to scientific problems—again, some of these have already been mentioned. The point is even more emphatic here than it was in §3b: in everything from astronomy to engineering, to navigation, to warfare, etc., the tools of analysis really worked. The result was what good old M. Kline calls "a virtual fusion of mathematics and vast areas of science."

The advantages of this fusion are more obvious than its dangers. Recall once more that a priceless feature of math is supposed to be the deductive, *a priori* truth of its theorems. Scientific truths are established empirically; they're inductive truths, and as such are subject to all the abstract early-morning uncertainties detailed in §1. Induction is, logically speaking, foundationless, whereas mathematical truths are built on the granite of axioms and rules of inference. All this has been discussed already, as have the connections between foundations and rigor, plus the thing in §3c's EMERGENCY GLOSSARY I about analysis trying (eventually) to inject more rigor into calculus.

The point: It's not enough that mathematical theories work; they're also supposed to be rigorously defined and proved in a way that meets the great Greek deductive standard. This is not what happened throughout much of the 1700s, though. It was really more like a stock-market bubble. And it looked great for a while. The "virtual fusion" in which mathematical discoveries enabled scientific advances which themselves motivated further math discoveries[3] created for math a situation that resembled a tree[4] with great lush proliferant systems of branches but no real roots. There were still no grounded, rigorous definitions of the differential, derivative, integral, limit, or convergent/divergent series. Not even of the function. There was constant controversy, and yet at the same time nobody seemed to care.[5] The fact that calculus's infinitesimals (and/or now the ∞-type limits toward which quantities could 'tend' without ever quite arriving) worked so well without any coherent foundation—this sort of infected the whole spirit of analysis. Without anyone explicitly saying so, math began to operate inductively.

It was in the areas of functions, differential equations, and trigonometric series that many of the 1700s' most significant advances and ghastly confusions arose. W/r/t our Story, it's going to be important to look at some of the specific math and science problems these concepts got used on. This in turn will require another relevant (though somewhat

[3] **IYI** meaning new results not only in established math but in whole new post-calculus fields, including differential equations, various kinds of infinite series, differential geometry, number theory, function theory, projective geometry, calculus of variations, continued fractions, and so on.

[4] **IYI** Notice we're back to the tree thing.

[5] *Vide* here not only d'Alembert's famous rationalization for not having a rigorous proof of the limits concept, "Just keep moving forward, and faith will come to you," but also this 1740ish pronouncement by A.-C. Clairaut (1713–1765, big math-physicist): "[It used to be that] geometry must, like logic, rely on formal reasoning in order to rebut the quibblers. But the tables have turned. All reasoning concerned with what common sense knows in advance, serves only to conceal the truth and to weary the reader and is today disregarded."

harder[6]) EMERGENCY GLOSSARY, to which you are again welcome to devote exactly as much time/attention as your background and interest warrant (but which you should probably at least skim and then be ready to flip back to later if difficulties *in situ* might warrant), etc.

EMERGENCY GLOSSARY II

—*Derivative* (*n.*) v. *Differential* (*n.*) These need to be distinguished even though they're so closely connected that a derivative is sometimes called a 'differential coefficient'. Recollect from E.G.I that a *derivative* is the rate of change of a function w/r/t the independent variable. In the case of a simple function like $y = f(x)$,[7] the derivative is $\frac{dy}{dx}$. What the individual dy and dx here are, though, are *differentials.* In something like $y = f(x)$, where x is the independent variable, the *differential* of x (that is, dx) is any arbitrary change in the value of x, in which case dy can be defined via $dy = f'(x)dx$, where $f'(x)$ is the derivative of $f(x)$. (Make sense? If $f'(x) = \frac{dy}{dx}$, then dy is pretty obviously $f'(x)dx$.)

An easy way to keep the two D-words straight is to remember that a derivative is literally the ratio of two

[6] Command Admission: Frankly, parts of E.G.II are going to be brutal, and on the whole this may be the hardest part of the entire booklet, and regrets are hereby conveyed. But it really is better to do it all here in one dense contextless chunk than to have to keep stopping and giving endless little defs. and glosses in the middle of describing people's actual work. It was tried both ways in drafts, and $\text{Evil}_1 < \text{Evil}_2$.

[7] **IYI** This universal symbolism for functions is courtesy of the prenominate L. Euler, the great towering figure of eighteenth-century math.

differentials—which is how many Leibnizians actually defined the derivative in the first place.

—*Partial Derivative* v. *Total Differential* is the pertinent distinction for 'functions of several variables,'[8] i.e. those with more than one independent variable. A *partial derivative* is the rate of change of a multivariable function w/r/t one of the relevant variables, the others being treated as constants—so generally a function will have as many partial derivatives as it has independent variables. A special symbol that Dr. G. called the 'dyslexic 6' is used for partial derivatives, as in e.g. the partial derivatives of the function-equation for the volume of a right cylinder, $V = \pi r^2 h$, which are $\frac{\partial V}{\partial r} = 2\pi rh$ and $\frac{\partial V}{\partial h} = \pi r^2$. A *total differential*, on the other hand, is the differential of a function with more than one independent variable—which usually amounts to saying it's the differential of the dependent variable. '∂'s get used for total differentials too. For a multivariable function like $z = f(x,y)$, z's total differential dz will be $\frac{\partial f}{\partial x} dx + \frac{\partial f}{\partial y} dy$.

These first two entries might seem excessively rarefied but are as a matter of fact required for

—*Differential Equations* (a–c (with (b) being rather expansive)), which are the #1 math tool for solving problems in physics, engineering, telemetry, automation, and all manner of hard science. You usually just start flirting with D.E.s at the end of freshman math; it's in Calc III that you find out how ubiquitous and difficult they really are.

[8] = what school calls Multivariable Calculus, which is essentially the math of surfaces in 3-space—$f(x, y)$—of solids in 4-space—$f(x, y, z)$—etc.

(a) In a broad sense, *differential equations* involve relationships between an independent x, a dependent y, and some derivative(s) of y with respect to x. D.E.s can be thought of either as integral calc on some sort of Class IV hallucinogen or (better[9]) as 'metafunctions,' meaning one level of abstraction up from regular functions—meaning in turn that if an ordinary function is a sort of machine where you plug certain numbers in and get other numbers out,[10] a differential equation is one where you plug certain functions in and get other functions out. The solution of a particular differential equation, then, is always some function, specifically one that can be substituted for the D.E.'s dependent variable to create what's known as an 'identity,' which is basically a mathematical tautology.

That may not have been too helpful. In more concrete[11] terms, a simple differential equation like $\frac{dy}{dx} = 3x^2 - 1$ has as its solution that function for which $3x^2 - 1$ is the derivative. This means what's now required is integration, i.e. finding just the function(s) that satisfies $\int(3x^2 - 1)\,dx$. If you've retained some freshman math, you'll probably see that $\int(3x^2 - 1)\,dx$ equals $f(x) = x^3 - x + C$ (with C being the infamous Constant of Integration[12]), which equation is the same as $y = x^3 - x + C$, which latter just so happens to be the *general*

[9] **IYI** The following was Dr. Goris's way of explaining differential equations, which turned out to be clearer and more significant than the formula-heavy way they get presented in college math.

[10] **IYI** Gorisian factoid: The original Japanese ideogram for 'function' meant, literally, 'number-box'.

[11] so to speak.

[12] long story—please just be aware there is such a thing.

solution of the differential equation $\frac{dy}{dx} = 3x^2 - 1$. This D.E.'s *particular solutions* will be those functions in which C takes some specific value, as in like $y = x^3 - x + 2$ and so on.

(b) Graphwise, because of C and the general/particular thing, differential equations tend to yield 'families of curves' as solutions. The equations' expansions, on the other hand, normally yield *sequences of functions*—and be advised that the move in analysis from sequences/series of quantities to sequences/series of functions ends up being crucial to the Story of ∞. Historically speaking, this move characterizes math's transition from the 1700s' Euler-type analysis to the more Cauchyesque kind of the early-to-mid-1800s. As was briefly mentioned in §3, Baron A.-L. Cauchy is credited with the first real attempt to rigorize analysis; he came up with a more sophisticated, convergence-based limit concept and was able to define continuity, infinitesimals, and even ∞ in terms of it.[13]

It was also Cauchy who first worked seriously on series of functions, in which the really crucial problems also involve convergence.*

***(QUICK EMBEDDED INTERPOLATION to (b)**

—*D.E.*(b) is about to get complicated. Besides being invited to flip back to E.G.I where appropriate, you are here apprised

[13] All this is right there in the first chapter of Cauchy's famous *Course d'analyse*, e.g. for $\frac{1}{\infty}$ and ∞: "A variable quantity becomes infinitely small when its numerical value decreases indefinitely in such a manner as to converge to the limit 0," and "A variable quantity becomes infinitely large when its numerical value increases indefinitely in such a manner as to converge to the limit ∞" (with, however, as M. Kline points out, "∞ mean[ing] not a fixed quantity but something indefinitely large").

of two related facts: (1) Convergence is (or at least looks) very different for sequences/series of functions than for sequences/series of quantities. For instance, what a convergent series of functions converges *to* is a certain function . . . or rather it's more precise to say that the *sum* of a convergent series of functions will always converge to a function.[14] (2) There are a whole roiling slew of connections between the concepts of continuity for functions and convergence for series of functions. Luckily, only a few of them concern us, but on the whole these connections go right to the troubled heart of nineteenth-century analysis, and some of the stuff gets extremely involved. An example is a famous mistake Cauchy made, which is presented here just as an indication of how bound up continuity and convergence are w/r/t functions. Cauchy held that if the sum of a sequence of continuous functions $c_0, c_1, c_2, \ldots$ converges everywhere on a certain interval to a function C, then function C is itself continuous on that interval. Why this was important, and wrong, will hopefully become clearer in a couple §s.

END Q.E.I. RETURN TO ¶2 of (b), IN PROGRESS)

Since you can't really do partial sums on series of functions, it becomes important to devise general tests for the convergence of these series/sequences. A pioneering general test called the Cauchy Convergence Condition (or '3C') of the 1820s holds that an infinite sequence $a_0, a_1, a_2, \ldots, a_n, \ldots$ converges (i.e., has a limit) if and only if the absolute value of

[14] Another advantage to thinking of differential equations as metafunctions is that, since sequences/series of functions are what D.E.s expand to, it makes sense that what, as it were, pops out the other end of such sequences/series is supposed to be a function.

$(a_{n+r} - a_n)$ is less than any specified quantity for every value of r, assuming n is sufficiently large.[15]

Final fact anent functions and convergence (which probably should have been included in the above Q.E.I.): To show that a function F is *representable* (= expandable) as an infinite series, you have to be able to verify that the series converges at all points to F. Which obviously can't be done by summing an infinite number of terms; you need an abstract proof. This, too, will be important when we get to Cantor's early work in analysis.

(c) One of the reasons differential equations are so hard in school is that there are many different types and subtypes of them, specified by all sorts of high-tech nomenclature—'order,' 'degree,' 'separability,' 'homogeneity,' 'linearity,' 'lag,' 'growth-'-v.-'decay factors,' etc. etc. For us, the most important distinction is between an *ordinary differential equation* and a *partial differential equation.* A *partial D.E.* involves more than one independent variable, and thus partial derivatives (hence the name), whereas an *ordinary D.E.* doesn't have any partial derivatives. Since most physical phenomena are complicated enough to require multivariable functions and partial derivatives, it is not surprising that the truly useful and significant differential equations are the partial

[15] **IYI** You might know or recall from college math that this is not at all like the general convergence test that's taught today. The reason is another semi-mistake of Cauchy's: it turns out that his original 3C can be rigorously proven only to be a *necessary* condition for convergence. A bona fide test for convergence also has to provide *sufficient* conditions,* for which it turns out you need a theory of real numbers, which won't be available until the 1870s.

* (q.v. §1c FN 14's rundown of necessary v. sufficient conditions)

ones. A couple other D.E. terms to know are *boundary conditions* and *initial conditions*, which have to do with specifying allowable values of y and/or whatever constants there are in the equation. The terms are germane because there are important connections between these conditions and the specification of certain Real Line intervals over which a function can range, which latter will be critical for the mid-1800s' Weierstrassian analysis coming up in §5e.

—*The Wave Equation* This is an especially famous and powerful partial D.E. Hugely influential in both pure and applied math, especially physics and engineering.[16] For our purposes, the relevant form is the 1D or 'nonLaPlacian' form of the Wave Equation, which (**IYI**) looks like:

$$\frac{\partial^2 y}{\partial x^2} = \frac{1}{c^2}\frac{\partial^2 y}{\partial t^2}$$

—*Trigonometric Series*, which probably should have gotten covered in E.G.I, are basically series whose terms are written as the sines and cosines[17] of various angles. The generic form is usually something like $\frac{a_0}{2} + \sum_{k=1}^{\infty}(a_k \cos kx + b_k \sin kx + \cdots)$. Trigonometric series play a major role in our Story, not only because they comprise Fourier Series as a subtype,[18] but also

[16] **IYI** Re the latter two, or in an astronomy class, you may have learned about the Wave Equation in association with *Bessel Functions*, which are particular solutions to the W.E. expressed in a special kind of 3D coordinate system.

[17] which of course are, strictly speaking, trigonometric *functions*, so you can see why trig series are a classic case of the series-of-functions stuff discussed above in —*Differential Equations* (b).

[18] It actually wouldn't hurt to flip back to E.G.I and check out —*Fourier Series* again, especially the stuff about F.S.s being expansions of periodic

because certain very important functions can be represented by both trig series and partial differential equations. The connections between partial D.E.s and trig series go right to the root of what a function is, it turns out; and in fact the modern math definition of a function—which differs from E.G.I's ad hoc def. in specifying that the association between each x and its $f(x)$ can be 100% arbitrary, with no rule or even explanation required[19]—is the result of exhaustive work on the relations between functions and their representations as series.

—*Uniform Convergence* & Associated Arcana (a–e). These items involve E.G.I's definitions of *interval* and *continuous function* as well as the stuff about the convergence of series of functions in the Cauchy part of —*Differential Equations* (b) just above. Besides being necessary for understanding certain big pre-Cantor results in later §s, this entry will afford you some idea of the truly vertiginous abstraction of nineteenth-century analysis.

(a) Core definition: A series of continuous functions of some x in some interval (p, q) is *uniformly convergent* if it converges for every value of x between p and q. There's also some boilerplate about 'remainders'[20] of the series being arbitrarily small that we can skip. The crux is that the sum of a

---

functions. This is because trig series themselves tend to be periodic, meaning they basically repeat the same wave over and over—w/ 'period' referring to the time required for one complete oscillation, and $y = \sin x$ (alias *the sine wave*) being a prototypical periodic function. (**IYI** If you happen to recall the term *oscillating series*, this is a totally different and unconnected thing and should be purged from memory for the remainder of this booklet.)

[19] long story, more or less unfolds over the next few §s.

[20] Don't ask.

uniformly convergent series will itself be a continuous function of x in the interval (p, q).

(b) Just as not all series are convergent, not all convergent series are uniformly convergent. Nor are all convergent series *monotonic*, which essentially means changing in the same direction all the time. Example of a monotonic decreasing series: the Dichotomous $\frac{1}{2}+\frac{1}{4}+\frac{1}{8}+\frac{1}{16}+\ldots$[21].

(c) Related to the thing about series converging in a given interval (p, q) is the matter of a function $f(x)$ being *sectionally continuous* in a given interval (p, q), which obtains when (p, q) can be divided into a finite number of subintervals, with $f(x)$ being continuous in each subinterval and having a finite limit at each lower $(= p)$ and upper $(= q)$ endpoint. (Note here that 'sectional' can also modify/specify monotonicness,[22] i.e. that some but not all monotonic series are *sectionally monotonic*—a bit of ephemera you'll need on board for one part of §5d.)

(d) For complicated reasons, if a function is sectionally continuous, it will have only a finite number of *discontinuities* in its relevant (p, q) interval. Here's a name that makes total sense: a *discontinuity* is simply a point[23] at which the function $f(x)$ is not continuous. Example: the semi-Fourierish $y = \dfrac{\sin(x-\alpha)}{1-\cos(x-\alpha)}$ will clearly have discontinuities at whatever

[21] **IYI** A *monotonic function*, on the other hand, is one whose first derivative doesn't change its +/− sign regardless whether the derivative's continuous or not. (Pretty sure we're not going to have to deal with m.f.s, although Weierstrassian analysis tends to require everything but the sink.)

[22] **IYI** Nominative form = 'monotony'? Surely not. Nothing on the math-noun in any sources. . . .

[23] here meaning a point in the domain of $f(x)$.

values of x make $\cos(x - \alpha)$ equal to 1. There are many different subspecies of discontinuities, which fact we will mostly ignore. Graphically, a discontinuity is a point at which a curve isn't smooth, i.e. where it jumps, or plummets, or there's maybe even a hole. Note also a bit of semantic finery: since the word 'discontinuity' can also refer, 2nd-Level-abstractly, to the general condition of something's being not-continuous, the term *exceptional point* is sometimes used to refer to a specific point at which there's a discontinuity. The nub here being that analysis tends to use 'discontinuity' and 'exceptional point' interchangeably.

(e) Last: Deceptively similar in English to 'uniform convergence' is *absolute convergence*, which mathwise is a totally different thing. A convergent infinite series S can have negative terms (e.g., the Grandi Series from §3a). If any/all of S's negative terms are made positive (that is, if only the *absolute values* of the terms are allowed) and S still converges, then it's *absolutely convergent*; otherwise it's *conditionally convergent.*

END E.G.II

§5b. Thrust of much of §5 so far: Floating around crucial but unmoored through the 1700s are ideas about functions, continuous functions, convergent functions, etc., with all their different respective defs. and properties undergoing constant change and refinement as analysis tackled various problems. As mentioned, and just as in the 1600s, a lot of these problems were scientific/physical. Here are some of the big ones of the eighteenth century: the behavior of flexible chains suspended from two points (a.k.a. 'catenary problems'), motions of a point along descending curves (= 'the brachistocrone'), elastic beams under tension, motions of a pendulum in resistant media, forms taken by a sail under wind pressure (= 'velaria'), orbits of planets w/r/t one another, caustic curves in optics, and fixed-compass movements on a sphere (= 'rhumb lines'). For our purposes, most important of all is the infamous Vibrating String Problem, which in some ways harks back to Pythagoras's discoveries about the diatonic scale in §2a. The general V.S.P. is: Given the length, initial position, and tension of a transversely pulled string, calculate its movements when it's released to start vibrating. These movements will be curves, which is also to say functions.

The reason the V.S.P. is often a mainstay of sophomore math is that it marks the first real application of partial differential equations to a physical problem. Here is some history. In the 1740s, J.•l.R. d'Alembert proposes what is basically the 1D Wave Equation as the correct representation of the V.S.P., yielding[24] the general solution $y = f(x + ct) + g(x - ct)$ where x is a point on a string of length π, y is the transverse displacement of x at time t, c is a constant,[25] and f and g are functions determined by the initial conditions. Where things get controversial is in the allowable scope of f and g. It turns out that d'Alembert's solution works only if the 'initial curve' of the string (that is, the way it's stretched at the start) is itself a periodic function. This puts a big restriction on the stretch, whereas of course for math and science you want maximally general solutions; and so major players like L. Euler, D. Bernoulli, J. L. LaGrange, P. S. LaPlace, and

[24] The following specifics are **IYI**.

[25] **IYI** This is the same c as in the Wave Equation—it's defined as 'the velocity of the propogation of the wave' or something close to that.

d'Alembert[26] all start arguing heatedly with one another about whether and how to let the string's initial stretch be any sort of curve/function at all and still make the Wave Equation apply. In brief, the consensus that finally emerges is that regardless what the string's initial shape is, its vibrational curves are going to be periodic functions, specifically sine waves. From which, for complicated reasons, it follows that no matter how the string is stretched at the outset—meaning any continuous curve at all—this curve will be representable by a trigonometric series.

Still in brief: A great many important discoveries about the nature and relations of functions, differential equations, and trigonometric series result from the opera of disagreement over the V.S.P. The one that's crucial to our Story is this idea that any continuous function[27] can be represented as a trig series. First Euler, then d'Alembert, J. L. LaGrange, and the aforementionedly quixotic A. C. Clairaut all start coming up with methods for representing 'arbitrary functions' as trig series. The trouble is that these methods are always derived and applied w/r/t some particular physical problem or other, the solution of which problem is then claimed to be the method's justification. Nobody's able to prove the

[26] **IYI** For some reason, this is a period in which nearly all the important mathematicians are French or Swiss. In the next century it will be the Germans who dominate. No good explanation for this in the literature. Maybe math is like geopolitics or pro sports, with different dynasties always developing and then fading, etc.

[27] which (again, and as Dr. G. himself always used to iterate and stress because he said if we didn't get this we would never understand how the V.S.P. and Wave Equation were connected to trig series) is the same thing as a curve.

function → trig series thing as an abstract theorem. This is all going on through about the end of the 1700s.

Now it's the early 1800s, during which time Cauchy and Norway's N. H. Abel[28] start doing significant work on series-convergence, which work ends up being even more significant in 1822. This is the year that the French Baron J.-B. J. Fourier[29] (1768–1830), working on problems in the conduction of heat in metals, demonstrates in his *Analytic Theory of Heat* that representability by trig series could actually be established for both continuous and *dis*continuous functions,[30] even for 'freely drawn' curves. Fourier's demonstration in *ATH* is too technical to get much into, but basically what he does is exploit the relation between the sum of a series and the integral of a function: he realizes that series-representability for wholly arbitrary functions requires ignoring the F.T.C. and defining integration geometrically[31] instead of just as the inverse of differentiation.

Since many classes teach Fourier Series without explaining where the math comes from, it's worth at least mentioning

[28] **IYI** 1802–29; joins E. Galois as the century's two great tragic prodigies; long, sad story; exerted (Abel did) an especially fertile posthumous influence on K. Weierstrass.

[29] **IYI** Curious addendum to FN 26's mystery: A lot of the preeminent French mathematicians of this era were also nobles—LaPlace a marquis, LaGrange a comte, Cauchy a baron, and so on—w/ at least some of these titles conferred by Napoleon I. Fourier's barony was, anyway; his own father was a tailor.

[30] Short version of story behind Fourier's discontinuous $f(x)$'s: Apparently, as a body takes on heat, its temperature gets distributed non-uniformly, meaning different spots have different temps at different times. It's the distribution of heat that Fourier's really interested in.

[31] i.e., as an area or sum of areas.

that Fourier starts out with a 2nd-order partial differential equation for the diffusion of heat in a 1D body, $\frac{\partial^2 y}{\partial x^2} = \frac{\partial y}{\partial t}$,[32] where y is the temperature of a point x at t, after which he uses a standard D.E.-technique called 'separation of variables,' plus the initial condition that $y = f(x)$, to derive that very special trig series now known as the Fourier Series, of which the relevant form is here presented as Exhibit 5b:

Exhibit 5b

$$f(x) = \frac{1}{2}a_0 + \sum_{n=1}^{\infty} (a_n \cos nx + b_n \sin nx) \quad \text{for } 0 \leq x \leq 2\pi$$

As it happens, this Fourier Series is actually very close to what D. Bernoulli had proposed as a solution to the V.S.P. back in the 1750s, except Fourier is able to calculate the series' coefficients[33] for every value between 0 and 2π. In the case of b_n, for example, $b_n = \frac{1}{\pi}\int_0^{2\pi} f(u) \sin nu \, du$—i.e., if you integrate term by term, the coefficients b_n will be $\frac{1}{\pi}$ times the area under the curve $(f(u)\sin nu)$ in the interval between $u = 0$ and $u = 2\pi$. With similar-type calculations for a_0 and a_n.

Eyeglazing or no, the point is that, via the formulae for these Fourier coefficients, every conceivable single-valued function—algebraic, transcendental, continuous, and even

[32] This partial D.E. is commonly known as the *Diffusion Equation*, which you might notice resembles the Wave Equation. It is just this resemblance that Fourier Series are able to account for, mathematically speaking.

[33] Since we pledged in E.G.I to try to steer clear of Fourier coefficients, the following five text-lines are classified **IYI**.

discontinuous[34]—becomes representable by a Fourier-type trig series on the interval $[0, 2\pi]$. There are all sorts of wonderful strengths and advantages to this technique (which Fourier developed primarily to give a general solution to the Diffusion Equation (which verily he did give)). One example: Understanding integration geometrically, and conceiving of a function in terms of its values rather than as just an analytic expression,[35] allows Fourier to consider functions' series-representability only over finite intervals, which is a major advance in flexibility for nineteenth-century analysis.

At the same time, though, Fourier Series are almost a rerun of early calculus in terms of the practical-efficacy-v.-deductive-rigor thing. Especially as refined by S. D. Poisson in the 1820s, Fourier Series become the #1 way to solve partial differential equations—which are, as mentioned, the golden keys to mathematical physics, dynamics, astronomy, etc.—and as such they pretty much revolutionize math and science all over again. But they are also foundationless; there is nothing like a rigorous theory of Fourier Series. In the words of one math-historian, Fourier's techniques "raised more questions than he was interested in answering or capable of solving." Which is both tactful and true: Fourier's *ATH* states but does not prove that a 'wholly arbitrary' function

[34] *Discontinuous function* is best thought of here as meaning that you can't express the function as a single '$y = f(x)$'-type equation—see one main-text ¶ down.

[35] *Analytic expression* basically means the '$y = f(x)$' thing. Another way of stating the text-clause is that Fourier interprets a function *denotatively*, i.e. as a set of specific correspondences between values, rather than *connotatively* as the name of the rule that generates the correspondences. Which, as we'll see in the next §, is a very modern way to think of functions.

can be represented by a series like Exhibit 5b; nor does it spell out what specific conditions a function has to satisfy to be so representable. Even more important, Fourier claims that his eponymous Series are always convergent in an interval regardless of what the function is or whether it's even expressible as a single '$y = f(x)$'; and while this has important implications for the theory of functions, there is no proof or even test for the 100%-convergence claim.

(**IYI** There was a similar problem involving *Fourier Integrals*, about which all we have to know is that they're special kinds of 'closed-form' solutions to partial differential equations which, again, Fourier claims work for any arbitrary functions, and which do indeed seem to—work, that is—being especially good for physics problems. But neither Fourier nor anyone else in the early 1820s can *prove* that Fourier Integrals work for all $f(x)$'s, in part because there's still deep confusion in math about how to define the integral . . . but anyway, the reason we're even mentioning the F.I. problem is that A.-L. Cauchy's work on it leads him to most of the quote-unquote rigorizing of analysis that he gets credit for, some of which rigor involves defining the integral as 'the limit of a sum' but most (= most of the rigor) concerns the convergence problems mentioned in (b) and its little **Q.E.I.** in the *—Differential Equations* part of E.G.II, specifically as those problems pertain to Fourier Series.[36])

There's another way to state the general difficulty. Fourier (rather like Leibniz and Bolzano) has an essentially geometric way of understanding things, and a penchant for geometrical demonstrations rather than formal proofs. In many respects, these are a holdover from classical calc and the

[36] There's really nothing to be done about the preceding sentence except apologize.

results-matter-more-than-proofs mentality of the 1700s. But such an approach is increasingly untenable now. Fourier's 1820s is also the decade when the first non-Euclidean geometries (based primarily around the discovery that the *Elements*'s Parallel Axiom[37] was dispensable) are discovered, and the idea that geometry could be any kind of fixed, univocal foundation for anything becomes officially absurd. A related issue is that mathematicians from Newton to Euler to C. F. Gauss had gotten into terrible paradoxical trouble using series without regard for convergence v. divergence,[38] and Fourier and Cauchy's emphases on convergence-in-intervals now help reveal just how sloppy analysis had been w/r/t series. The overall result is the start of a correction in analysis's stock-market bubble; or, as M. Kline has it, "[M]athematicians began to be concerned about the looseness in the concepts and proofs of the vast branches of analysis." It's in the 1820s that we start getting pronouncements like Cauchy's "It would be a serious error to think that one can find certainty only in geometrical demonstrations or in the testimony of the senses" and Abel's "There are so very few theorems in advanced analysis that have been demonstrated in a logically sound way. Everywhere one finds this wretched method of concluding from the special to the general,"[39] which latter became as famous a sound bite for nineteenth-century retrenchment as d'Alembert's "Just keep moving forward" had been for the 1700s' laissez faire.

[37] **IYI** q.v. §1d.

[38] **IYI** Recall e.g. Euler's $\frac{1}{1-x}$ canard in §3a, or for that matter the whole Grandi Series thing.

[39] Important to notice: Abel's final gerund phrase is just a roundabout way of saying 'induction'.

Overall point: Along with the fall of Euclid, it's Fourier Series' arbitrary-function- and convergence-issues which prompt the era's mathematicians to realize that atomic concepts like 'derivative,' 'integral,' 'limit,' 'function,' 'continuous,' and 'convergent' had to be rigorously defined, w/ 'rigorously' here meaning basing analysis on formal proofs and arithmetical reasoning instead of on geometry, intuition, or induction from specific problems.

Except 'arithmetical' in turn meant the real-number system, which at this time was itself still an ungrounded mess. There were, for example, hideous problems with negative numbers—Euler was convinced that negatives were actually $> \infty$, i.e. that they ought to be way out to the right on the Number Line; and as late as the 1840s A. De Morgan held that negatives were just as 'imaginary' as $\sqrt{-1}$; and let's not even talk about the snafus over complex numbers. The worst trouble, though, was that the root concept of 'real number' was itself unclear because irrationals were still undefined. If you can't coherently define numbers like $\sqrt{2}$ or $\sqrt{3}$, you can't prove any of the basic arithmetic laws for them, e.g. that $\sqrt{2} \times \sqrt{3} = \sqrt{2 \times 3}$.[40] This is not good, rigorwise. There's a certain amount of valuable sidework in this period on transcendental v. algebraic irrationals,[41] but for the most part

[40] **IYI** Nor can you demonstrate this equivalence by calculation, since of course '$\sqrt{2}$' and '$\sqrt{3}$' both represent infinite decimals. In the nomenclature of analysis, you can't prove that product of the sums of these two decimals' infinite series converge to the sum of $\sqrt{6}$.

[41] **IYI** Said work is by J. Liouville and C. Hermite (more Frenchmen). Re the whole algebraic-v.-transcendental-irrational thing, see or resummon §3a FN 15. There's also some more on Liouville's big proof later on in §7c.

the real-number system on which everybody's laboring to ground analysis is itself dangling in midair, logically speaking.

§5c. SOFT-NEWS INTERPOLATION, PLACED HERE ANTE REM BECAUSE THIS IS THE LAST PLACE TO DO IT WITHOUT DISRUPTING THE JUGGERNAUT-LIKE MOMENTUM OF THE PRE-CANTOR MATHEMATICAL CONTEXT

There are several extant photos of G. F. L. P. Cantor in books, at least one of which can hopefully be appropriated and reproduced here someplace. He is a completely average-looking bourgeois German from the era of starched collars and fire-hazard beards. (Note, in family photos, the waistcoat and pocketwatch w/ prominent fob, the wife's plaits and bustle, the sternly serene or abstracted expression of the standard Victorian male. In the U.S. he'd have had a hat.)

In or about 1940, special National Socialist historians of mathematics 'discover' that G. F. L. P. Cantor had been a foundling, born and discovered on a German ship on its way to the port of St. Petersburg, parents unknown. Which is utter fiction. The Reich was apparently worried that Cantor might have been Jewish; by then he was regarded as one of Germany's greatest intellectuals ever. The foundling story still circulates sometimes—it fits some of our own templates as well as the Nazis'. Another big one is that Cantor derived many of his most famous proofs about ∞ while in an asylum, which is also hooey. Cantor's first hospitalization was in 1884, when he was 39; most of his important work had already been done by then. He wasn't hospitalized again until 1899. It was in the last 20 years of his life that he was in and out of places all the time. He died in the Halle *Nervenklinic* 6 January 1918.

This photo is missing the plaits and fob, but you get the idea.

The Cantor family home on *Handelstrasse* was at least briefly occupied during WWII. There's no evidence that the Nazis knew whose house it had been. Still, the major portions of Cantor's literary estate were evidently lost or burned. Most of

Here is the pencil sketch referred to a couple of ¶s down—it's not entirely clear to me why they put it here.

what's left is at the *Akademie der Wissenschaften* in Göttingen and is available for perusal behind glass. Family letters, genealogies, etc. There are also still a few of Cantor's letter-books, which were what literate people then used to draft letters before copying them carefully out to send. Plus there were other mathematicians he wrote to who kept his letters. These are the primary sources.

Here is a quotation from J. W. Dauben, the dean of U.S. Cantor-scholars: "Too little information has been preserved to allow any detailed assessment of Cantor's personality, which leaves the historian to say either nothing on the subject, or to conjecture as best he can." Much of the published conjecture concerns

Cantor's father, Mr. Georg W. Cantor, with the big modern issue being whether "Georg Woldemar had a thoroughly deleterious and ruinous effect upon his son's psychological health"[42] or whether Georg W. was actually "a sensitive and gifted man, who loved his children deeply and wanted them to live happy, successful, and rewarding lives." Either way, it's for sure that there were exactly two businessmen who had a profound effect on G. F. L. P. Cantor's life, one being his father and the other Prof. Leopold Kronecker, who starts looming large in §6.

Mr. and Mrs. Georg W. have six kids; Georg Jr. is the first. The whole extended family is artistic and high-functioning: several relatives are classical violinists or showing painters; a great-great-uncle had been director of the Vienna Conservatory and the teacher of the virtuoso Joseph Joachim; a great-uncle had been Tolstoy's law professor at Kazan University. G. F. L. P. Cantor's date of birth is 3 March 1845. A Pisces. Something of a violin prodigy as a child. No one knows why he quit, but after a classical quartet in college there's no more mention of the violin. Also a good natural artist. A pretty extraordinary pencil sketch from childhood survives; it's famous because of a lefthanded and indisputably creepy "Proclamation" that Georg W. conferred on it, which proclamation itself survives because (also creepily) Georg Jr. kept it on his person all his life:

> Whereas Georg Ferd. Louis Phil. Cantor has not spent years in the study of drawing according to the ancient

[42] Another exemplary quotation along these lines is E. T. Bell's "Had Cantor been brought up as an independent human being he would never have acquired the timid deference to men of established reputation which made his life wretched."

> models; and whereas this is his first work and in this difficult art a perfected technique is only achieved after great diligence; and whereas, furthermore, until now he has greatly neglected this beautiful art; the thanks of the nation—I mean of the family—is unanimously voted to him for this first effort, which already shows great promise.

All sources agree that Georg W. personally supervised his children's religious development in something of a hardass way. Notice how easy it is to view this through today's lens as oppressive or neurogenic, when in fact it might have been just SOP for the time and place. It's hard to tell. Likewise the fact that Cantor Sr. "took a special interest in [Georg's] education and was careful to direct his personal and his intellectual development."

The family moves from St. Petersburg to western Germany when Georg is 11. The reason, according to historians, is Georg W.'s "poor health," which in the 1850s was code for TB; the relocation is analogous to moving from Chicago to Scottsdale. They live mainly in Frankfurt, on the Rhine. Georg boards at prep schools in Darmstadt and Wiesbaden. As seems to happen with most great mathematicians, Cantor's analytical genius gets discovered in his early teens; ecstatic letters from his math teachers are still there to be seen at the *Akadamie.* The story's standard version is that Georg W. wants Georg Jr.'s gifts put to practical uses and tries to force the boy into engineering, that Georg burns to do pure math and has to hector and beg, etc., and that when Georg W. finally accedes it's in a way that puts great pressure on his fragile son to Achieve and Excel. Again, it's not clear whether this was quite true or/and how unique a father-son gestalt it was.[43]

Cantor does his undergrad work in Zurich and then gets the German equivalent of an M.S. and Ph.D. at U. Berlin, which at that time is sort of the MIT of Europe. His teachers at Berlin include E. E. Kummer, L. Kronecker, and K. Weierstrass. It's Kronecker who is Cantor's dissertation advisor and his real mentor and champion in the department. The irony of this will also start emerging in §6.

§5d. Now we're back in the 1820s with Fourier and trig series and all the challenges and opportunities attendant thereon. If §5b's discussion of Fourier Series and E.G.II's —*Differential Equations*'s interpolated stuff on the connections between continuity and convergence were halfway lucid, it will not surprise you to learn that in *Analytic Theory of Heat* Fourier supplies the first modern definition of convergence, as well as introduces the vital idea of convergence

[43] **IYI** Another exchange of letters is often cited, w/ all weirdnesses and emphases *sic*—

> Georg W → Georg Jr.:
> . . . I close with these words: Your father, or rather your parents and all other members of the family both in Germany and in Russia and in Denmark have their eyes on *you* as the eldest, and expect you to be nothing *less* than a Theodor Schaeffer [teacher of G. C. Jr.] and, God willing, later perhaps a *shining star* on the horizon of science.

> Georg Jr. → Georg W.:
> . . . Now I am happy when I see that it will no longer distress you if I follow my own feelings in this decision. I hope that you will be proud of me one day, dear Father, for my soul, my entire being lives in my calling; whatever one wants and is able to do, whatever it is toward which an unknown, secret voice [?!] calls him, *that* he will carry through to success!

in an interval. But that (again) Fourier fails to give a rigorous proof, or even to spell out convergence-criteria that would make such a proof possible. So the thing is now we're talking about convergence.

It's Bolzano[44] and more famously Cauchy who did the first salient work on conditions/tests for convergence. As has been mentioned in a couple places already, a lot of Cauchy's results were valuable, but he also made weird errors that caused further problems. Though a lot of his work was on series of functions, for instance, Cauchy chose to define 'limit' in terms of variables instead of functions. Or an even better example here is the way that he tried to establish a total identity between convergence and continuity in terms of sequences/series of functions. Recall[45] that Cauchy's claim is that if a sequence of continuous functions converges in interval I to a function C, then C itself is continuous on I—which it turns out is not true unless the convergent sequence is *uniformly* convergent. What happens in this case is that N. Abel, in proving Cauchy's error[46] and refining the theorem, develops what's now known as Abel's Uniform-Convergence Test,[47] just

[44] **IYI** Suffice to say that Bolzano's definitions and results on continuity in §3c can, with only minor adjustments, be extended to the concept of series-convergence.

[45] **IYI** from E.G.II's —*D.E.*(b)'s Q.E.I.

[46] **IYI** This is 1826. Abel's specific counterexample is the series $\sin x - \frac{\sin 2x}{2} + \frac{\sin 3x}{3} - \cdots$, which happens to be the Fourier Series expansion of $y = \frac{x}{2}$ in the interval $-\pi < x < \pi$, and verily is convergent, but whose sum is discontinuous for $x = \pi(2n + 1)$ where n is an integer.

[47] **IYI** which A.U.C.T. is still used today, and you might have had it in school; and if you didn't, here it is so you can see what this sort of thing really looks like: Assume $c_n(x)$ is a sequence of functions in some interval

as all sorts of other criteria and conditions for various kinds of convergence of various kinds of trig and polynomial series are also getting developed as various nineteenth-century mathematicians struggle to clean up other mathematicians' mistakes and/or to solve problems in better ways.

One of the era's most important cleanup mathematicians is G. P. L. Dirichlet (1805–1859), a friend of Fourier's whose 1829 "*Sur la convergence des séries trigonométriques*" did much to clarify and rigorize what came to be known as the General Convergence Problem of Fourier Series. There are several important advances in this paper, like that Dirichlet is the first to discover and distinguish absolute v. conditional convergence; plus he disproves Cauchy's contention that a monotonic decreasing series is the same as a convergent series. Most important, though, is that in "*Sur la convergence*" Dirichlet establishes and proves the first set of sufficient conditions[48] for a Fourier Series converging to its original $f(x)$.

This last result is pertinent and deserves some detail. Dirichlet uses a periodic $f(x)$ on the interval $[-\pi, \pi]$ and

---

$[a, b]$. If (1) the sequence can be rewritten as $c_n(x) = a_n f_n(x)$, and if (2) the series Σa_n is uniformly convergent, and if (3) $f_n(x)$ is a monotonic decreasing sequence such that $f_{n+1}(x) \leq f_n(x)$ for all n, and if (4) $f_n(x)$ is bounded in $[a, b]$, then (5) for all x in $[a, b]$, the whole series Σc_n is uniformly convergent. (If (2)'s requirement that Σa_n be uniformly convergent looks odd/circular, be advised that it's a common trick in pure math to take a property of something simple or easily proved (Σa_n being far and away the simpler part of (1)'s decomposition of $c_n(x)$) and to use it to prove that the same property holds for a more complicated entity. In fact, this very trick is at the heart of *mathematical induction*, a 100% reputable proof-technique that enters play in §7.)

[48] **IYI** Please see or recall §5a's FN 15.

basically proves that if (1) its Fourier Series is sectionally continuous and thus has only a finite number of discontinuities[49] in the interval, and if (2) the Series is sectionally monotonic, then (3) the Series will always converge to the $f(x)$, even if that function requires more than one sort of '$y = f(x)$' expression to represent it in the interval.

There's one additional requirement, which Dirichlet also proves. The function $f(x)$ has to be *integrable*—that is, $\int_{-\pi}^{\pi} f(x)dx$ has to be finite—basically because the relevant Series' Fourier coefficients are calculated as integrals and these need to be 'well-defined' (long story). As an example of a function that isn't integrable and so can't be represented by a well-defined Fourier Series, Dirichlet cooks up a pathological $f(x)$ whose values equal the constant c when x is rational but equal the constant d (w/ $d \neq c$) when x is irrational, which function is indeed not integrable. It's largely this pathological $f(x)$ that leads Dirichlet, eight years later,[50] to give the definition of 'function' that's still used in modern math: "y is a function of x when to each value of x in a given interval there corresponds a unique value of y." The big thing is that the correspondence can be completely arbitrary; it doesn't matter whether y's dependence on x accords with any particular

[49] **IYI** again, a.k.a. exceptional points.

[50] **IYI** = 1837, in a paper whose rather lovely-sounding title is "*Über die Darstellung ganz willkürlicher Functionen durch Sinus- und Cosinusreihen,*" which translates roughly to "On the Representation of Wholly Arbitrary Functions by Sine and Cosine Series." (Impressively, mathematicians of this era seem able to write in both French and German, depending on what journal they're submitting a paper to.)

rule, or even whether it can be expressed mathematically.[51] Strange as this sounds, it's actually 100% rigorous for math purposes, since arbitrariness yields maximum generality, a.k.a. abstraction. (Dirichlet's definition also happens to be very close to the Bolzano-Cantorian idea of one-to-one correspondence between two sets of real numbers, except of course neither 'set' nor 'real number' has been defined in math yet.)

In his 1837 article, Dirichlet is also able to show that you could relax the monotonic requirement (2) and even allow a greater number of discontinuities (1) in his 1829 proof and still guarantee a Fourier Series' convergence to its integrable $f(x)$ . . . so long as the number of discontinuities in (1) remains finite. This is still not the same as proving Fourier Series–convergence for *any* arbitrary $f(x)$, though—especially when you're trying to Fourierize the hairy and often wildly discontinuous functions of pure analysis and number theory.[52] W/r/t these complex functions, the question Dirichlet is never able to answer is whether criterion (1) of his proof could be relaxed to allow *infinitely*[53] many discontinuities in the interval and still constitute a sufficient condition for convergence.

Now G. F. B. Riemann (1826–1866, that colossus of pure math who revolutionized everything from functions to number

[51] meaning that now in analysis a function is really neither a thing nor a procedure but rather just a set of correspondences between a domain and a range.

[52] We should mention that what Fourier, Cauchy, and Dirichlet usually worked with were functions in mathematical physics, which tend to be comparatively simple and well-behaved.

[53] **IYI** Right here is the first bit of real foreshadowing w/r/t how hardcore analysis spawns transfinite math.

theory to geometry,[54] and is Cantor's only serious competition for Mathematician of the Century) enters the picture, albeit briefly and in a mostly transitional role. Twenty years

[54] One hesitates to get into this, but in fact the Riemannian version of non-Euclidean geometry—sometimes a.k.a. 'general differential geometry,' and dating from 1854 (a very big year for Riemann)—constitutes a whole other vector of explanation for why rigorous theories of real numbers and ∞ become necessary in the 1800s. It's slightly tangential, and brutally abstract, and will be mostly pretermitted except for right here.

Boiled way down, Riemannian geometry involves (a) Gauss's complex plane (i.e., a Cartesian grid that has real numbers as one axis and complex numbers as the other axis) and (b) something called a Riemann Sphere, which can be thought of as basically a 2D Euclidean plane curved into a ball and set atop the complex plane. This footnote isn't technically **IYI**, but feel free to stop anytime. What connects Riemannian geometry with the aforementioned projective geometry of Desargues is that every point on the Riemann Sphere has a 'shadow' on the complex plane; and the trigonometric relations created by these shadows turn out to be fecund as hell, ∞-wise. For example, a line on the complex plane is the shadow of something called a Great Circle on the Riemann Sphere, meaning a circle whose circumference goes through the R.S.'s north pole, which pole is defined, literally, as 'a point at ∞'. In fact the entire Riemann Sphere is definable as 'the complex plane with a point at ∞,' which entity is also known as the Extended Complex Plane. 0 is the Riemann Sphere's south pole, and ∞ and 0 are by differential-geometric definition inversely related (because taking the inverse of a number on the complex plane is equivalent to flipping the Riemann Sphere over—long story). So that in Riemannian geometry, '$0 = \frac{1}{\infty}$' and '$\infty = \frac{1}{0}$' are not only legal; they're theorems.

We'll stop the specific discussion here and hope it makes some kind of general sense. Big overall thing to know: It is not an accident that the symbol for the Extended Complex Plane is '$\mathbb{C}_\infty$' whereas G. Cantor's most famous symbol for the set of all real numbers (a.k.a. the Continuum, which as we'll see is basically the 2nd mathematical order of ∞) is '**c**'. There are all kinds of fascinating connections between Riemannian geometry and Cantorian set theory, most of which are unfortunately beyond what we're set up to talk about.

older than Cantor, Riemann is a student of Dirichlet's at U. Berlin, as well as a friend of R. Dedekind.[55] In a seminal 1854 paper,[56] Riemann attacks the General Convergence Problem of Fourier Series in a whole new way. Focusing on the proviso that an $f(x)$ has to be integrable in order to be representable as a Fourier Series, he derives general conditions that any function must satisfy to have an integral, conditions that end up being important both for analysis's theory of integration and for series-convergence. In essence, 'general conditions that any function must satisfy' means necessary conditions, whereas you'll recall that Dirichlet's 1829 proof had involved sufficient conditions. It's by inverting his teacher's approach and concentrating on necessary conditions for convergence that Riemann solves Dirichlet's big problem: he constructs a function that has infinitely many discontinuities in every interval but is nevertheless integrable and 100% convergent at every point.[57] A consequence of this

[55] **IYI** By this time, all the major players in the genesis of transfinite math are alive, and most are mathematically active. In 1854, Riemann is 28, Dedekind 23; Weierstrass is 39 and L. Kronecker 31. The not-yet-mentioned E. H. Heine is 33. Cantor is 9 and playing the violin under the beady eye of Georg W.

[56] **IYI** This paper's long title starts with "*Über die Darstellbarkeit* . . ." (= "On the Representability . . . "). It was actually more of a second Ph.D. dissertation than a pro monograph (long story), and its mss. got handed around among mathematicians until Dedekind finally arranged to have it published after Riemann died.

[57] **IYI** For the stouthearted and/or hale of background, or if you maybe just want to revel in some more of analysis's symbology: Riemann derives this $f(x)$ by taking a standard trig series and integrating each term twice, yielding $f(x) = C + C'x + \frac{a_0}{2}x^2 - \sum_{r=1}^{\infty} \frac{a_r \cos rx + b_r \sin rx}{r^2}$, whereupon he's able to prove that this trig series is convergent so long

result is something known as Riemann's Localization Theorem, which states that the convergence of a trig series at a point depends all and only on the behavior of its relevant $f(x)$ in some arbitrarily small neighborhood[58] of that point. And the 'arbitrarily small neighborhood' thing is what finally validates Fourier and Dirichlet's claims about series representing wholly arbitrary functions: via the Localization Theorem, even highly discontinuous or pathological functions can be represented by trig series—and, if integrable, by Fourier Series.

As is SOP, though, at the same time Riemann's work is answering earlier questions it's also raising new ones; and it's ultimately these questions that make his 1854 paper so important. Example: An implication of the Localization Theorem is that two different integrable functions can be represented by the same trigonometric series even if they differ at a large but finite number of points—is there any way to get the same result for two functions that differ at an infinity of points? Other important ones: Exactly what properties of trig series allow them to be convergent even with a ∞ of exceptional points in each interval? and How exactly are the concepts of $f(x)$-continuity, intervals, and neighborhoods connected to the theory of trig series? and Is every trig series a Fourier Series (i.e., does every trig series converge to an integrable function)? and If more than one function can be represented by the same

as $\displaystyle \lim_{\alpha \to 0, \beta \to 0} \frac{f(x+\alpha+\beta) - f(x+\alpha-\beta) - f(x-\alpha+\beta) + f(x-\alpha-\beta)}{4\alpha\beta}$

behaves in certain special ways (which ways, again, we don't have the conceptual ordnance to talk about, but they're not dubious or strange, just really technical).

[58] **IYI** as defined in E.G.I.

trig series, is the reverse true, or does each unique $f(x)$ have only one unique trig-series representation?[59]

After Reimann's paper, the next big order of business for pure mathematicians is coming up with the techniques needed to solve the problems it had posed, w/ the special challenge being to ground those techniques in rigorous foundations instead of the inductive or faith-based intuition that had marked so much past analysis. N.B. that the emphasis on foundations/rigor is partly because Riemann's wholly abstract functions have finally moved Fourier Series out of the applied-math realm of physics and into higher math per se. But plus of course we're now in the 1850s, and the need for rigor (as discussed re the 1820s in §5b) is even more generally urgent. In real time it's taken several decades, but the boom of justification-by-results is now giving way entirely to the more contractionist, prove-as-you-go economy of what math-historians all call the Arithmetization of Analysis.

§5e. The key figure at this point is Karl Weierstrass (1815–1897), who can now dramatically be revealed as one of the heroes of B. Russell's thing about Zeno's Paradoxes in §2a. Here's the rest of that quotation[60]:

> From [Zeno] to our own day, the finest intellects of each generation in turn attacked the problems, but achieved, broadly speaking, nothing. In our own time, however, three men—Weierstrass, Dedekind, and Cantor—have

[59] **IYI** More foreshadowing: The first and last of these questions will be what G. F. L. P. Cantor's own early work tries to answer, and it's this work that leads him into ∞ per se.

[60] **IYI** The preceeding parts appear on pp. 48 and 52.

not merely advanced the problems, but have completely solved them. The solutions, for those acquainted with mathematics, are so clear as to leave no longer the slightest doubt or difficulty. . . . Of the three problems, that of the infinitesimal was solved by Weierstrass; the solution of the other two was begun by Dedekind and definitively accomplished by Cantor.

Weierstrass is not directly in the Fourier-Cauchy-Dirichlet-Riemann academic line—up until his 40s, he's obscure in the same way Bolzano was. His early career is spent teaching high school in West Prussia (not exactly a hub),[61] and he's said to have been literally too poor to afford the postage for submitting work to journals. He finally starts publishing in the late 1850s, and sets math on its collective ear, and gets hired by prestigious U. Berlin as a prof—it's all a long and kind of romantic story. (**IYI** Weierstrass is also conspicuous among mathematicians for being physically large, a gifted athlete, an inveterate partier and blowoff in college, indifferent to music (most mathematicians are fiends for music), and a cheery, non-neurotic, gregarious, wholly good and much-loved fellow. He's also widely regarded as the greatest math teacher of the century, even though he never published his lectures or even let his students take notes.[62])

[61] **IYI** N.B., however, that German technical *Hochschulen* were extremely hardass institutions by today's standards, and calc and basic analysis were required parts of the curriculum. (Teacher salaries legendarily low, though.)

[62] The grad student who did take covert notes and is the main reason why Weierstrass's later function-theory got known is one G. Mittag-Leffler (1846–1927), who later started the famous journal *Acta Mathematica* and

The specific reason we're now talking about Weierstrass is that it is mostly his discoveries that enable math to attack the questions that Dirichlet and Riemann's work on the G.C.P.F.S. had raised. So much so that q.v., from math-historian I. Grattan-Guinness, "[T]he history of mathematical analysis during the last third of the 19th century is in notable measure the story of mathematicians applying Weierstrassian techniques to Riemannian problems." The real inspiration behind these techniques is not Fourier or Riemann but the tangentially aforementioned N. H. Abel (Weierstrass being a huge Abel fan), specifically an innovation called elliptic functions that Abel had derived c. 1825 from *elliptic integrals*—which latter, to make a long story short, emerge in calculating the arc-length of an ellipse and are a big deal in both pure and applied math.[63] Weierstrass's first significant work (back in W. Prussia, by candlelight, in between grading quizzes) involves the power-series expansions of elliptic functions, which leads him into problems regarding the convergence of

---

published G. Cantor's work on ∞ when most other math journals still thought it was insane. Historically, Mittag-Leffler is regarded as Cantor's 2nd most important penpal, after R. Dedekind.

[63] **IYI** For the most part, elliptic integrals are generalizations of inverse trig functions; they tend to show up in all sorts of physics problems, from electromagnetism to gravity. If you see them in a math class, it's usually in conjunction with A. M. Legendre (another of the early 1800s' Frenchmen), who was to elliptic integrals what Fourier was to trig series, and developed 'Legendre's standard elliptic integrals of the 1st, 2nd, and 3rd kinds.' (**IYI$_2$** If, incidentally, you know that G. Riemann also did a lot with E.I.s and certain associated integrals in the Calculus of Variations, be advised that we're not getting into any of this.)

power series,[64] and thence to convergence, continuity, and functions in general.

The reason Russell lauds him w/r/t the problem of infinitesimals is the same reason Weierstrass gets top billing in the Arithmetization of Analysis. He is the first to give a wholly rigorous and metaphysically untainted theory of limits. Because it's important, and underlies the way most of us are now taught calc in school, let's at least quickly observe that Weierstrass's definition of limits replaces Abel/Bolzano/Cauchy's natural-language terms like 'approaches a limit' and 'becomes less than any given quantity' with the little epsilon and delta and the '| |' brackets of absolute value. A great fringe benefit of Weierstrass's theory is that it characterizes limits and continuity in such a way that either can be defined in terms of the other. See for example his definition of *continuous function*, which is still the industry standard 150 years later[65]: $f(x)$ is *continuous* at some point x_n if and only if, for any positive number ε, there exists a positive δ such that for any x in the interval $|x - x_n| < \delta$, $|f(x) - f(x_n)| < \varepsilon$.*

*** QUICK EMBEDDED SEMI-IYI INTERPOLATION**

Please skip the following three ¶s if and only if Weierstrass's definition already makes total sense to you.

Since the prenominate def. is not **IYI**, we need to countenance the possibility that it may be <100% clear why it's such a big deal. We could talk about how the definition is specifically a rigorization of Cauchy's *Cours d'analyse*'s "$f(x)$ will be

[64] **IYI** Pretty sure E.G.I mentioned that Fourier Series can be thought of as power-series-sums.

[65] For reasons that will become apparent, this high-tech def. is *not* **IYI**.

a *continuous* function of the variable if the numerical value of the difference $f(x + \alpha) - f(x)$ decreases indefinitely with that of α"—Weierstrass's coup is that he comes up with a rigorous, wholly arithmetical substitute for the murky "decreases indefinitely". But none of that is going to mean much if we can't see how Weierstrass's new definition actually works, which in turn requires parsing the dense technical syntax in which it's laid out for mathematicians. Its hyper-abstract language can make the def. seem either totally trivial (e.g., since ε and δ are not defined as having any direct relation to each other, isn't it obvious that you can pick a δ for any ε you want?) or totally mystifying (e.g., how can you determine what $|x - x_n|$ equals if you don't know what x is?). Or at least such were our class' initial complaints to Dr. Goris, who handled them in his usual unforgettable way, more or less as follows[66]:

First off, recall from E.G.I how *continuous function* means that tiny little changes in the function's independent variable (x) yield only tiny little changes in the dependent variable ($f(x)$, alias y). The 'tiny little changes' thing is what's really important here: it clues you in that differences like $|x - x_n|$ in the def. are going to be conceived as really small. Respecting x_n and x: x_n is a particular point, namely the point for which we're going to evaluate the continuity of the function; and x is, technically, any point at all in the function's interval, although given what *continuous function* means it's better to

[66] Command Admission: What follows is going to be a lot less formal/rigorous than Dr. G.'s own gloss of the def. We're aiming for an explanation that will make maximum sense in terms of the vocab and concepts developed in the booklet thus far.

think of x as any point that's 'sufficiently close' to x_n. This is because the whole point of Weierstrass's def. is to let us verify that a tiny little difference between x and x_n will yield only a tiny little difference between $f(x)$ and $f(x_n)$. The engine of this verification is the positive numbers ε and δ, and the easy way to understand these two numbers and their relation is in terms of a game. Here's the game: you pick any positive ε you want, no matter how small, and I try to find a positive δ that will make the conjunction $(|x - x_n| < \delta)$ & $(|(f(x) - f(x_n)| < \varepsilon)$ true[67]; and if I can find such a δ for any ε you pick, then the $f(x)$ is continuous at x_n, and if I can't it isn't.

As did Dr. G., let's do an example for a function that isn't continuous, so that we can see how not just any old δ will do for a given ε. The function's defined thus: $f(x) = 1$ if $x \neq 0$, and $f(x) = 0$ if $x = 0$. We're going to evaluate this $f(x)$'s continuity at the point x_n where $x_n = 0$. The game is that you can pick any positive value you want for ε, and let's say you pick $\varepsilon = \frac{1}{2}$. So now I have to find a positive δ such that $(|x - x_n| < \delta)$ & $(|f(x) - f(x_n)| < \frac{1}{2})$ will be true. But now please either flip or think back to §1c FN 14's rule that a logical conjunction is true only when *both* its pre-'&' term and

[67] 100% technically speaking, Weierstrass's def.'s "for any . . . , there exists . . . such that" is really laying out an entailment-relation in *first-order predicate calculus*, which is a more involved kind of logic that uses quantifiers like '$\forall$' and '$\exists$'. Except for one or two toss-offs in §7, though, this booklet does not get into predicate calculus, so here we're symbolizing the relation between the definition's ε and δ as a logical conjunction. For our illustrative-proof purposes this will work just as well; the relevant truth-values end up the same.

its post-'&' term are true. And now look at our example's post-ampersand term, '$|f(x) - f(x_n)| < \frac{1}{2}$'. By the definition of our function, we know that $f(x_n) = 0$; and by the same definition, we know that any other x besides x_n yields an $f(x)$ of 1 (because x_n is the only point at which $x = 0$). So where $x_n = 0$, $|f(x) - f(x_n)|$ is always going to equal 1, which is obviously greater than $\frac{1}{2}$. No matter what δ I might pick for $|x - x_n|$ to be less than, $|f(x) - f(x_n)|$ is never going to be $< \frac{1}{2}$ when $x_n = 0$. Since the second term will always be false, the conjunction $(|x - x_n| < \delta)$ & $(|f(x) - f(x_n)| < \frac{1}{2})$ is going to be false regardless what δ is. The upshot of which is that there is *not* a positive δ for $\varepsilon = \frac{1}{2}$ here, so the definition's criterion, "for any positive number ε, there exists a positive δ such that . . . ," is not satisfied. So the function isn't continuous at x_n (which we knew to begin with, but the whole object was to apply Weierstrass's def. to a clear case). Game over.

END Q. E. S.- IYI I. RETURN TO MAIN §5e DISCUSSION, IN PROGRESS

One bit of possible puzzlement not covered in the **INTERPOLATION**: What Weierstrass has defined supra is just continuity-at-a-point—but since you can choose any point in a given interval to be x_n, an $f(x)$ can obviously be defined as continuous in an interval if it's continuous at each and every x_n in the interval. So a general definition of continuity for functions drops right out of the initial def. And—this is what rocked the math world—so does the definition of *limit*: if you take Weierstrass's original def., then $f(x)$ can be defined as having a limit L at x_n if you can replace $f(x_n)$ with L and still find a δ for any ε such that $(|x - x_n| < \delta)$ & $(|f(x) - L| < \varepsilon)$ is true.

There's a reason this all looks so hideously abstract: it *is* hideously abstract. Yet this very abstractness is what makes Weierstrass's the cleanest, clearest theory of continuity/limits anyone had ever come up with.[68] There's no natural-language fuzz; the def. uses nothing but real numbers and primitive operators like '$-$' and '$<$'. Because it's so clean and abstract and arithmetical, the theory also enables Weierstrass to define convergence v. uniform convergence v. absolute convergence rigorously, to provide bona fide tests for same,[69] and to prove a number of things about continuity and trig-series-convergence that nobody'd been able to nail down. Pertinent examples here include (1) his proof that a series of continuous functions[70] can converge to a discontinuous function, and (2) his aforementioned disproof of the theory that continuity = differentiability, which disproof he demonstrates by deriving a function that is 100% continuous but has a derivative at no point. If you're curious, this is the $f(x)$ given by $\sum_{r=0}^{\infty} b^r \cos(a^r \pi x)$ where a is odd, $0 < b < 1$, and $2ab > 2 + 3\pi$—which incidentally if you graph this $f(x)$ you get a curve that's got no tangent at all. You'll remember from §3c that B. P. Bolzano had already come up with a similar function (which there's no evidence Weierstrass knew about), but there's a crucial difference. Bolzano had merely proffered his example, whereas thanks to his purely formal definitions

[68] **IYI** Factoid: Such is the significance of Weierstrass's technique here that a great many subsequent proofs in analysis and number-theory use the "For any ε, there exists a . . . " device, for which the sexy term is *epsilontic proof*.

[69] **IYI** His test for uniform convergence, known as Weierstrass's M-Test, still gets taught in analysis classes.

[70] e.g., a Fourier Series.

Weierstrass has the ammo to really *prove* that his own $f(x)$ is both continuous and not differentiable. Another major advance: In Weierstrassian analysis, the concrete example is always coincident to the abstract general proof.

As also mentioned in §3c, Bolzano and Weierstrass get joint credit for a big theorem about the limits of a continuous function,[71] a theorem that's apposite here in part because it conceives of an infinite series/sequence of real numbers as an infinite set of Real Line points (and the first kinds of infinite sets that will interest G. Cantor are these *point sets*). You'll probably want to recall the entries for —*Interval* and —*Limits* v. *Bounds* from E.G.I here. Formally speaking, the Bolzano-Weierstrass Theorem holds that every bounded infinite set of points contains at least one *limit point*, which is a point x_n whereby every interval around x_n contains a ∞ of members of the set.[72] It may not immediately look it, but the B.W.T. is a true ball of fire. For instance, it turns out to yield a potent, infinitesimal-free antidote to §4b's no-next-instant paradox (which cruncher is itself consequent to the mindbending density of the Real Line's ∞ of points).

[71] **IYI** Historical factoid: Weierstrass's contribution to this theorem is really part of his unsuccessful attempt to define irrational numbers—q.v. §6a below.

[72] In essence what this does is recast the concept of convergence-to-a-limit in terms of point sets and R.L.-intervals. Example from S. Lavine: "1 is a limit point of the set $\{0, \frac{1}{2}, \frac{3}{4}, \frac{7}{8}, \ldots\}$. Intuitively one sees that the members of the set crowd against 1." Granted, this isn't a totally rigorous def. of *limit point*, but it's certainly the upshot. Note here also that the limit point of an infinite *sequence* is a point such that every interval around it has an infinite number of terms of the sequence—meaning it works pretty much the same way, which will become relevant in §7a.

If you're up for this, here's how it works. The B.W.T. actually comprises two theorems, one of which is Bolzano's from c. 1830 and states that, given a closed interval $[a, b]$, a continuous $f(x)$ in $[a, b]$ that is positive for some value of x and negative for some other value of x must equal 0 for some value of x.[73] Geometrically, we can make sense of this by seeing that a continuous curve that goes from someplace above the x-axis to someplace below the x-axis has to actually hit the x-axis at some point. Rigorwise, Bolzano's problem was that proving this theorem required him to prove that every bounded set of values/points must have a least upper bound,[74] which latter proof foundered on the absence of coherent theories of limits and real numbers. So once again Bolzano could really only propose his theorem and demonstrate geometrically that it was true; he couldn't formally prove it. Twenty years later, though, K. Weierstrass uses his rigorous theory of limits to prove Bolzano's 'Least Upper Bound Lemma' as part of his own *Extreme Values Theorem*, which is the other big part of the B.W.T. It's the Extreme Values Theorem (= if an $f(x)$ is continuous in $[a, b]$, then there

[73] **IYI** A direct consequence of Bolzano's theorem is something usually called the *Intermediate Values Theorem*, which is a staple of function-theory and says in essence that if an $f(x)$ is continuous in $[a, b]$ such that $f(a) = A$ and $f(b) = B$, then $f(x)$ takes all possible values between A and B. If we define the continuous function as $f(x) = 2x$, $[a, b]$ as $[0, 1]$ and $[A, B]$ as $[0, 2]$, then you can see that a prime instance of the I.V.T. is Bolzano's §3c proof about the one-to-one correspondence between $[0, 1]$ and $[0, 2]$. This in turn might make it clearer why Bolzano's theorem requires a theory of real numbers, since 'all possible values between A and B' is not going to be a quantity we can verify by counting.

[74] **IYI** also in E.G.I's thing on —*Limits* v. *Bounds.*

must be at least one point in $[a, b]$ where the $f(x)$ has its absolute maximum value M and another point in $[a, b]$ where $f(x)$ has its absolute minimum value m), together with Weierstrass's high-powered definition of continuity, that provides a mathematical way out of the no-next-instant crevasse. To wit: Since time is clearly a continuously flowing function,[75] we can assume a finite interval $[t_1, t_2] > 0$ between any two instants t_1 and t_2, and now, thanks to the E.V.T., prove that there is a point in $[t_1, t_2]$ where the time-function has its absolute minimum value m, and therefore that this t_m will be, mathematically speaking, the *very next instant* after t_1.

From this result, you can probably see how the Extreme Values Theorem could be deployed against Zeno's Dichotomy itself (since $\frac{1}{2^n}$ is a prototypical continuous function). In strict Weierstrassian analysis, though, the E.V.T. isn't even necessary, since the arithmetical theory of limits allows us to explain—meaningfully—why the convergent series $\frac{1}{2^1} + \frac{1}{2^2} + \frac{1}{2^3} + \cdots + \frac{1}{2^n}$ has 1 as its sum.

§5e(1). INTERPOLATION ON WEIERSTRASSIAN ANALYSIS AND ZENO

You'll remember that we've already tried invoking $\frac{a}{1-r}$ and various other formulaic solutions to the Dichotomy, only to find that they do not "state clearly the difficulties involved," much less explain how you can ever cross the street. These prior tries can all now be more or less ignored—though it wouldn't hurt to recall or review §2c FN 35's thing about

[75] **IYI** Indeed, so far as the real world goes, it's the paradigm of such a function.

how decimals are representations of convergent series. Plus of course the last few pages of §5e just above. So then here's a Weierstrassian analysis–type response to the Dichotomy.[76]

What the Dichotomy's really talking about is a certain rational number s (= the width of the street, the length of the arc from lap to nose), which number Zeno is inviting us to approximate by a convergent power series of other rational numbers s_n, where n itself stands for the infinite sequence 1, 2, 3, ... That may look opaque in the abstract, but lots of rational numbers can be approximated the same way. The rational s $\frac{2}{3}$, for instance, can be approximated by the following convergent s_n: $\frac{6}{10^1} + \frac{6}{10^2} + \frac{6}{10^3} + \cdots + \frac{6}{10^n}$. The Dichotomy's only slightly trickier than the $\frac{2}{3}$ thing—and some of this trickiness can be mitigated by talking about §2b's more abstract, 'revised' Dichotomy, where there's no time or motion but simply some quantity s that's halved, half that half halved, half that half-half halved, etc., until the smallest portions start equaling $\frac{1}{2^n}$ where n is arbitrarily large.[77] Adding up the portions derived this way gives us the Dichotomy's s_n as the following

[76] What follows is, again, very informal, and customized to work w/r/t the math and logic concepts we've set up thus far. (**IYI** A 100% rigorous response would deploy the Weierstrassian def. for the limit of an infinite sequence, which is a somewhat different form of epsilontic proof that we've opted not to spend another INTERPOLATION unpacking. For our purposes, the limit-of-function def. will work just as well.)

[77] meaning n 'goes to' ∞ just like the integral sequence 1, 2, 3, ... (**IYI** Did we mention back in §2 how uncannily close the revised Dichotomy is to Eudoxus's own Exhaustion Property?)

convergent power series: $s_n = \frac{1}{2^1} + \frac{1}{2^2} + \frac{1}{2^3} + \cdots + \frac{1}{2^n}$. And the thing to see is that this s_n is an approximation of 1 in just the same way that .99999 . . . is an approximation of 1. That is, the sum of s_n differs from 1 only by $\frac{1}{2^n}$, and this difference will become arbitrarily small as n increases indefinitely.

Granted, the Cauchyesque 'arbitrarily' and 'indefinitely' here seem vague and unsatisfying—and Zeno wants them to; he wants us to be stymied by the fact that there are an infinite number of terms in '$\frac{1}{2^1} + \frac{1}{2^2} + \frac{1}{2^3} + \cdots$' and that in the real world you can never actually finish adding them up—in which case now Weierstrass to the rescue.

A couple preliminaries will make it easier to see how this works. First, be apprised that the index n also functions as the *ordinal number*[78] of any given term in the series s_n—i.e., that $\frac{1}{2^1}$ is the 1st term, $\frac{1}{2^4}$ is the 4th term, $\frac{1}{2^{47}}$ is the 47th term, etc. Notice, too, that the difference between any two successive terms of s_n gets smaller and smaller the farther out in the series you get. With 'difference' in that last clause being representable as a distance on the Number Line. Meaning we're talking about intervals.

[78] As you may already know, the adjective here comes from 'order,' and *ordinal number* means the 1st number, the 2nd number, the 3rd number, etc., as opposed to the *cardinal numbers* 1, 2, 3, etc. In other words, ordinality concerns *where* the number is in a given sequence rather than *what* it is. The cardinal-v.-ordinal distinction turns out to be mammothly important in Cantorian set theory, to foreshadow which see for instance B. Russell's "In this theory [of ∞], it is necessary to treat separately of cardinals and ordinals, which are far more diverse in their properties when they are transfinite than when they are finite."

So here we go. We know that the sum of the Dichotomy's s_n will differ from 1 only by $\frac{1}{2^n}$ where $n = 1, 2, 3, \ldots$. To prove that 1 is actually the sum of s_n, we have to prove that 1 is the limit of the function $(1 - \frac{1}{2^n})$ where $n = 1, 2, 3, \ldots$.[79] We prove this via Weierstrass's '$f(x)$ has a limit L at point x_n if and only if, for any positive number ε, there exists a positive number δ such that for any x in the interval $|x - x_n| < \delta$, $|f(x) - L| < \varepsilon$.' Here L and x_n are both 1, $f(x)$ is $(1 - \frac{1}{2^n})$, and x can be whatever we want—the simplest way to do the proof is to let $x =$ the point $\frac{1}{2^n}$. And remember (assuming you did read §5e's INTERPOLATION) that the def.'s odd syntax really means that we need to be able to find a δ for any ε such that the logical conjunction $(|x - x_n| < \delta)$ & $(|f(x) - L| < \varepsilon)$ is true. Then the only other preliminary is making sure to remember what the absolute-value signs mean: $|1 - 10|$ and $|10 - 1|$ both $= 9$, and $|f(x) - L|$ and $|L - f(x)|$ are also equivalent. Weierstrass uses absolute values because we're talking about intervals on the Number Line, i.e. about the numerical distance between different points, which is the same from either direction. Here, what the '| |'s let us do is in effect switch the two different subtractions' terms around, which will make the actual N.L.–intervals we're talking about easier to represent clearly.

Sorry about all the verbiage—this is really easier than we're making it sound. So: To prove that 1 is the limit of the Dichotomous function $(1 - \frac{1}{2^n})$, we have to find, for any

[79] **IYI** This is, of course, the same as proving that 0 is the limit of $\frac{1}{2^n}$ for $n = 1, 2, 3, \ldots$. We're doing the function $(1 - \frac{1}{2^n})$ for maximum perspicuity w/r/t the Dichotomy.

positive number ε such that the interval between 1 and $(1 - \frac{1}{2^n})$ is $< \varepsilon$, a positive δ such that $((1 - \frac{1}{2^n}) < \delta)$ & $(1 - (1 - \frac{1}{2^n}) < \varepsilon)$ is true. Which it turns out is not hard to do. W/r/t ε and the conjunction's second term, the situation is sort of the reverse of the INTERPOLATION's example: $(1 - (1 - \frac{1}{2^n}) < \varepsilon)$ is never going to be false no matter what positive value you choose for ε. Like say you decide on $\varepsilon = .001$. You can then make $(1 - (1 - \frac{1}{2^n}) < \varepsilon)$ true by letting n equal 10 (as it will in s_n's tenth term), in which case $\frac{1}{2^n} = \frac{1}{1{,}024}$, in which case $(1 - \frac{1}{2^n}) = \frac{1{,}023}{1{,}024}$, in which case $(1 - (1 - \frac{1}{2^n})) = (\frac{1{,}024}{1{,}024} - \frac{1{,}023}{1{,}024}) = \frac{1}{1{,}024}$, which is the same as .0009765, which verily is $< .001$. The point is that no matter what tiny little ε you pick, you can adjust the value of n so that $(1 - (1 - \frac{1}{2^n}))$ comes out to less than ε.

So the conjunction's second term's always going to be true. Now all we have to worry about is the conjunction's first term and finding a δ that will make $(1 - (\frac{1}{2^n}) < \delta)$ true for a given ε. We obviously do not have the same carte blanche w/r/t picking δ that we did with picking ε, because our choice of ε determines the value we assign to n and thus to $\frac{1}{2^n}$, and it's $(1 - \frac{1}{2^n})$ that δ's got to be greater than. But it turns out that δ's dependent relation to ε makes it simple to find a relevant δ. Let's look again at the example where $\varepsilon = .001$ and so $n = 10$ and $\frac{1}{2^n} = \frac{1}{1{,}024}$. Here we need a positive δ such that $(1 - \frac{1}{1{,}024}) < \delta$. In this particular case, $\delta = 1$ will work fine . . . and in fact it turns out that $\delta = 1$ will work for *every* possible value of ε. You can probably see why this is. It's because all ε's possible values must, by definition, be positive. Even though ε can get smaller and smaller—and the smaller ε gets, the larger n has to get in order to make $(1 - (1 - \frac{1}{2^n}))$ less than ε, and the larger n gets, the closer $(1 - \frac{1}{2^n})$ will get to 1—still, the requirement that ε be > 0 ensures that $(1 - \frac{1}{2^n})$ will always be <1. And since, no matter what positive ε you pick, $\delta = 1$ ensures that the conjunction $((1 - \frac{1}{2^n}) < 1)$ & $(1 - (1 - \frac{1}{2^n}) < \varepsilon)$ will be true, then the def.'s primary criterion, "for any positive number ε, there exists a positive δ," is satisfied. Hence $\lim_{n \to \infty}(1 - \frac{1}{2^n}) = 1$, hence 1 is the sum of s_n. Hence you really can cross the street.

The Dichotomy's central confusion is now laid bare: the task of moving from point A to point B involves not a ∞ of necessary subtasks, but rather a single task whose '1' can be validly approximated by a convergent infinite series. It is the mechanics of this approximation that Weierstrassian analysis is able to explain—meaning *really* explain, 100% arithmetically, without infinitesimals, analogies, or any of the natural-language ambiguity that Zeno'd thrived on. It is not impoverished to say that, after Weierstrass, the Dichotomy becomes just another Word Problem.

A coda, though. The proof we just did is unusually detailed. Rarely, in an undergrad classroom, will proofs that $\lim_{n \to \infty}(1 - \frac{1}{2^n}) = 1$ or $\lim_{n \to \infty}(\frac{1}{2^n}) = 0$ actually get into Weierstrass's intervals or ε/δ. These are now the as it were hidden foundation of the limits concept, its deductive justification; the concept itself still gets expressed in natural-language terms like 'indefinitely' and 'tends to'. Which is probably fine. Instead of any more screeds about technical v. genuine answers, let's merely observe that a standard math-class solution to the Dichotomy will basically stop after the 2nd paragraph of §5e(1). That is, a math class will set up the convergent series $s_n = \frac{1}{2^1} + \frac{1}{2^2} + \frac{1}{2^3} + \cdots + \frac{1}{2^n}$, point out that the difference between s_n and 1 is $\frac{1}{2^n}$, demonstrate that this difference becomes arbitrarily small as n increases indefinitely, and teach you that the right way to handle a series like this is "by saying that *the sum* $\mathrm{s_n}$ *approaches the limit* 1 *as* n *tends to infinity*, and by writing $1 = \frac{1}{2} + \frac{1}{2^2} + \frac{1}{2^3} + \frac{1}{2^4} + \cdots$." The material in quotes is from an actual math text, revised ed. © 1996, emphases *sic*. The same text continues:

> This "equation" does not mean that we actually have to add infinitely many terms; it is only an abbreviated expression for the fact that 1 is the limit of the finite sum s_n as n *tends to* infinity (by no means *is* infinity). The infinite enters only in the unending *procedure* and not as an actual *quantity*.

Let us assert, in a very calm and low-key way, that if you think you can detect a whiff of Aristotelianism in all these urgent assurances that ∞ does not have to be dealt with "as an actual *quantity*" in problems like the Dichotomy, you are not smelling things. Nor is this just a matter of the formulaic way math is now taught to undergrads. It's deeper than that, and older. A remarkable thing about nineteenth-century analysis and function-theory was that the more sophisticated they got, the more uncannily their treatment of ∞ came to resemble

Aristotle's hoary old 'potentiality' concept. And the very acme of this sophistication, and of this resemblance, is Weierstrassian analysis.

If, however, by some additional chance you've noticed that the Dichotomy and its putative classroom solution concern only rational numbers and the N.L.—not to mention that our own proof's exemplary ε's and δ's in §5e have also all been rational—then you're in a position to appreciate a sort of lovely irony.

§6a. Ultimately more important than any specific result is the spirit in which Weierstrass redefines limits and continuous functions. Meaning his commitment to foundational rigor, apodictic hygiene, etc. Weierstrass's own Arithmetization of Analysis is 100% literal. It aims not only to eliminate geometrical concepts and inductive intuition from proofs but to base all of post-calc math on the real-number system in the same way arithmetic's so based. But the real-number system means the Real Line, which as we've seen is not exactly short of ∞-type crevasses.

There are various ways math-historians put this, as in e.g. Kline: "It was Weierstrass who first pointed out that to establish carefully the properties of continuous functions he needed the theory of the arithmetic continuum," or Bell: "The irrationals which give us the concepts of limits and continuity, from which analysis springs, must be referred back by irrefrangible reasoning to the integers." The upshot is the irony promised at §5's end: Weierstrass's rigorous limits-concept, which appeared to have finally and coherently eliminated the need for ∞- and $\frac{1}{\infty}$-type quantities from analysis, turns out itself to require a clear, rigorous theory of real numbers, meaning surds and the continuum of the Real Line. Re which see math-philosopher S. Lavine's "And that theory promptly reintroduced the infinite into analysis. The old infinity of infinitesimal and infinite numbers was simply replaced by the new infinity of infinitely large collections."

The truth is that there are several different interrelated reasons why irrationals/reals become now a front-burner problem. One, as just mentioned, is foundational. Another involves applications: it turns out that the Weierstrassian ε's and δ's that show up in real-world problems' limits are often irrational, making it much harder to establish the δ-for-any-ε thing. Plus there's the aforementioned G.C.P.F.S., the loss of faith in Euclidean axioms, etc. There's also the fact that math's new emphasis on rigor and formal coherence highlights a logical problem in the way surds had been handled ever since the Divine Brotherhood first encountered them in §2c. That encounter was, as we saw, geometrical—meaning incommensurable magnitudes, $\sqrt{2}$, Eudoxian ratios, etc.—and working definitions of surds in math had been geometrical ever since. This practice was, in the pellucid terms of B. Russell,

> highly illogical; for if the application of numbers to [geometric] space is to yield anything but tautologies, the numbers applied must be independently defined; and if none but a geometrical definition were possible, there would be, properly speaking, no such arithmetical entities as the definition pretended to define.

Which argument would take too long to unpack all the way, but you can see the general drift.

So what happens is that in the 1860s and '70s various mathematicians start trying to work out rigorous theories of irrationals/reals. Big names here include W. R. Hamilton, H. Kossak, K. Weierstrass, F. Lindemann, H. C. R. Méray, G. Cantor, H. E. Heine, and R. Dedekind. Guess which ones we're interested in.

First, apparently working off ideas he'd outlined in his U. Berlin lectures, some of Weierstrass's students and followers try to use his key definitions to define an irrational number as, in essence, the limit of a particular kind of infinite series of rational numbers. The def.'s technique is convulsantly technical, but fortunately we don't even have to go into it because it turns out the theory doesn't work; it's circular.[1] This is because Weierstrass's irrational limit cannot exist, logically speaking, until there's a definition of irrational numbers. With 'exist' here meaning precisely what Russell means just above when he says "there would be no such arithmetical entities as the [geometric] definition pretended to define." You can't coherently use the concept of an irrational number in the definition of 'irrational number' any more than you can coherently define 'black' as 'the color of a black dog'. So, in brief, the Weierstrassians' efforts never really got anywhere.

G. Cantor's own theory of real numbers arises in the context of his work on something called the Uniqueness Theorem for trig series, and there are good reasons for waiting a little bit to talk about it.

The most potent, significant, and strange-looking scheme for defining irrationals is that of J. W. R. ('Richard') Dedekind, 1831–1916, who's a dozen years younger than Weierstrass but very much like him. An affable, well-adjusted person who

[1] The mathematician who first points this out: Prof. G. F. L. P. Cantor.

spends most of his life teaching at technical *Hochschulen* in Brunswick and Zurich. A career bachelor[2] who lives with his sister. Dedekind survives to such an old age and is so well liked that he shows up all over modern math: student of Dirichlet and Gauss at Göttingen, editor of Dirichlet's *Zahlentheorie*, lifelong friend of Riemann, semi-Weierstrassian, early collaborator with L. Kronecker on algebraic geometry, collaborator and friend of G. Cantor—and Cantor was not easy to be friends with. A favorite story among math-historians has Dedekind living so incredibly long that Teubner's famous *Mathematical Calendar* kept announcing his death on a certain day in 1899, until Dedekind finally one year wrote the editor that he was alive and had moreover spent the day in question "in stimulating conversation on 'system and theory' with my luncheon guest and honored friend Georg Cantor of Halle."

The pub.-date of Dedekind's famous "Continuity and Irrational Numbers"[3] is 1872. It's partly a response to Cantor's own def. of irrationals, which had appeared as part of a big paper on the Uniqueness Theorem earlier that year. But evidently Dedekind had his basic theory in place as early as

[2] **IYI** Odd factoid: Almost all history's great philosophers never married. Heidegger's the only real exception. The great mathematicians are nuptially split about 50/50, still way below the civilian average. No cogent explanation on record; feel free to hypothesize.

[3] **IYI** = "*Stetigkeit und irrationale Zahlen*," whose English translation is in Dedekind's *Essays on the Theory of Numbers*—q.v. the Bibliography. (Let's note here also that, despite its profundity, Dedekind's paper is clear and accessible and rarely requires anything more than high-school math. In this it's unlike Cantor's stuff, which tends to be near-Medusean in its language and symbolism.)

1860; like Weierstrass (and unlike Cantor), he's not very ambitious about publishing his stuff. Dedekind's motivation, again just like Weierstrass's, came from teaching high-school calc: he'd gotten more and more uneasy about using undefined geometrical concepts to define limits and continuity. Instead of focusing on the arithmetization of limits, though, Dedekind goes as it were deeper, to the root problem that had occupied Zeno, the D.B.P., Eudoxus, Aristotle, and Bolzano, and had haunted analysis since the Fundamental Theorem. That problem: How to derive a 100% arithmetical theory of what calculus was supposed to be concerned with—pure continuity, as in motion and continuous geometric constructs like lines and areas and volumes[4]—but had never defined with enough clarity and rigor to make its proofs truly sound.

The entity Dedekind chooses to emblemize arithmetical continuity is the good old Real Line—which technically we should still call the Number Line, since only after the establishment of what's called the *Cantor-Dedekind Axiom* does it become rigorously OK to talk about the Real Line. So call it the Number Line (Dedekind just calls it *L*), and recollect from §2c that it's ordered and infinitely dense and extended, and that you can use it to map the rational numbers by assigning to each rational a unique point on the Line. And yet that the whole reason analysis needs more than the rational numbers is that the Number Line, like all lines, is continuous in a way that the set of all rationals is not. The way Dedekind

[4] Weierstrass, on the other hand, was concerned primarily with continuity as it pertains to *functions*, which ultimately depends on arithmetical continuity (as Kline points out at §6a's start) but is still a different kettle.

puts this harks all the way back to the Greeks—"Of the greatest importance, however, is the fact that in the straight line *L* there are infinitely many points which correspond to no rational number"—citing as an example the good old Unit Square's diagonal. So his strategy is simply to explain what it is about *L* that makes it continuous and gap-free when the infinitely dense set of rational numbers is not. Of course, he and we and everyone else by then all know it's the irrational numbers, but no one's been able to define them directly. So Dedekind's going to play Socrates here and act as if he's never even heard of irrationals and simply ask: In what exactly does *L*'s continuity inhere?[5] It's his answer to this question that makes the Real Line a mathematical reality and establishes, with Cantor, that the set of all real numbers composes the Continuum.

§6b. INTERPOLATION

Known today as the Dedekind Cut, R.D.'s device for constructing the Real Line is extremely ingenious and strange, and before we succumb to its intricacies it's worth pointing out that in a deep sense it is Dedekind's willingness to treat of the actual infinite that enables his proof to go through. As has

[5] In case it seems circular or question-begging for Dedekind to be using the geometrical Number Line for a geometry-free theory of continuity, be advised that the N.L. is just an illustrative device, one that later on in "C. and I.R." he'll drop in favor of "any ordered system" of numbers. Even in introducing the Line, Dedekind makes it a point to say "[I]t will be necessary to bring out clearly the corresponding purely arithmetic properties in order to avoid even the appearance as if arithmetic were in need of ideas foreign to it."

been mentioned in §5 and elsewhere, one of the reasons the limits concept was so welcome in analysis was that it accorded nicely with the old idea of the potential infinite—the idea that ∞ is something you can 'approach' without ever actually having to get there seems almost right out of the *Metaphysics.* Math's tacit embrace of Aristotle's distinction gets made explicit in an oft-quoted statement of C. F. Gauss (yes: Dedekind's Gauss) c. 1830:

> I protest against the use of an infinite quantity as an actual entity; this is never allowed in mathematics. The infinite is only a manner of speaking, in which one properly speaks of limits to which certain ratios can come as near as desired, while others are permitted to increase without bound.

And so of course it's ironic that only a couple decades after Weierstrass rids limits of the last shadowy bits of ∞ you start seeing top mathematicians who not only embrace the actual infinite but use it in proofs. Dedekind is one of these. And it's not like he gets sold on the coherence of infinite sets by Cantor or Bolzano or anyone else. Dedekind is, as defined way back in §2a, a Platonist. He clearly believes that mathematical reality is not so much empirical as cognitive:

> In speaking of arithmetic (algebra, analysis) as a part of logic I mean to imply that I consider the number-concept entirely independent of the notions or intuitions of space and time, that I consider it an immediate result from the laws of thought.

Or maybe it's better to call him a phenomenologist, since to Dedekind the distinction between math realities as created v.

as discovered doesn't much seem to matter: "Numbers are free creations of the human mind; they serve as a means of apprehending more easily and more sharply the difference of things."

Or look at this. In a companion essay to "Continuity and I.N." that's usually translated as "The Nature and Meaning of Numbers,"[6] Dedekind evinces a remarkable proof for his "Theorem 66. There exist infinite systems," which runs thus: "My own realm of thoughts,[7] i.e., the totality *S* of all things which can be objects of my thought, is infinite. For if *s* signifies an element of *S*, then the thought *s'*, that *s* can be an object of my thought, is itself an element of *S*, . . ." and so on, meaning that the infinite series ([*s*] + [*s* is an object of thought] + ['*s* is an object of thought' is an object of thought] + · · ·) exists in the *Gedankenwelt*, which entails that the *Gedankenwelt* is itself infinite. With respect to this proof, notice (a) how closely it resembles the various Zeno-like VIRs back in §2a, and (b) how easily we could object that the proof establishes only that Dedekind's *Gedankenwelt* is 'potentially infinite' (and in precisely Aristotle's sense of the term), since nobody can ever actually sit down and think a whole infinite series of $(s + s' + s'')$-type thoughts—i.e., the series is a total abstraction.

The point is that Dedekind's proof in "Nature and Meaning" isn't going to convince anybody who's not already disposed to

[6] **IYI** This 1880s paper consists entirely of 171 Theorems and Proofs + 1 "Final Remark". You're hereby spared the German title. The English version is the other half of the *Essays* book mentioned in FN 3 supra.

[7] = German *Gedankenwelt*, literally 'thought-world'.

admit the existence[8] of actually infinite systems/series/sets . . . which both Dedekind and Cantor are. In spades. Unlike Dedekind, though, Cantor tends to represent himself as sort of dragged kicking and screaming into actual ∞s, as in e.g.:

> The idea of considering the infinitely large not only in the form of the unlimitedly increasing magnitude and in the closely related form of convergent infinite series . . . but to also fix it mathematically by numbers in the definite form of the completed infinite was logically forced upon me, almost against my will since it was contrary to traditions which I had come to cherish in the course of many years of scientific effort and investigations.

Part of this difference in presentation is that G. Cantor was a slicker rhetorician than R. Dedekind, and part is that he had to be—Cantor was on the front lines of math's battle over ∞ in a way that Dedekind never was.

One other thing to keep in mind, though, is that Cantor's transfinite math will end up totally undercutting Aristotelian objections like the above (b) to Dedekind's proof, since Cantor's theory will constitute direct evidence that actually-infinite sets can be understood and manipulated, truly *handled* by the human intellect, just as velocity and acceleration are handled by calculus. So one thing to appreciate up front is that, however abstract infinite systems are, after Cantor they are most definitely *not* abstract in the nonreal/unreal way that unicorns are.

END INTERP.

[8] meaning mathematically, although in the proof it's not clear whether Dedekind's "infinite system" *S* is represented as a strictly mathematical entity or as a more general Platonic-Formish one—which is another problem with his argument.

§6c. Back to Dedekind's opening gambit of asking what distinguishes the continuity of the Number Line *L*. As it happens, Galileo, Leibniz, and Bolzano had all tried positing that the continuity of *L* was really based on the infinite density of its constituent points—i.e., on the fact that between any two of the N.L.'s points you can always derive a third point. As we've seen, though, the same feature applies to all rationals of the form $\frac{p}{q}$,[9] and since (1) every rational number can be put into the form $\frac{p}{q}$,[10] and (2) we already know that the set of

[9] **IYI** You might recall that §2e's $q + (\frac{p-q}{2})$ proof of the 3rd-point thing involved distances on the Number Line, which might look a bit circular in the present context, in which case here's a 100% arithmetical formula for finding the medial value. Take any two successive rational numbers expressed as fractions, like say $\frac{41}{77}$ and $\frac{42}{77}$, and double all four terms, automatically allowing integral space in the numerators—$\frac{82}{154}$ and $\frac{84}{154}$—in which to insert the medial $\frac{83}{154}$.

[10] **IYI** An algorithm that people who find this sort of thing fun will find fun: Any rational number expressed as a terminating or periodic decimal can be transposed into $\frac{p}{q}$ form by (a) multiplying the decimal by 10^n, where n = the number of digits in the decimal's basic period (e.g., the period of .11111 . . . is 1; the period of 876.9567567567567 . . . is 3), then (b) subtracting the original decimal from the quantity in (a), then (c) dividing the result by $(10^n - 1)$, then (d) reducing the result by eliminating any common factors. Example: $x = 1.24242424\ldots$; so $n = 2$; so $10^n = 100$.

$$\begin{array}{r} 100x = 124.242424\ldots \\ -\ x = 1.242424\ldots \\ \hline 99x = 123 \end{array}$$

So $x = \frac{123}{99}$, which reduces to $x = \frac{41}{33}$.

all rationals isn't continuous, Dedekind dismisses the idea that *L*'s continuity inheres in any kind of density or 'hang-togetherness': "By vague remarks upon the unbroken connection in the smallest parts [of *L*] obviously nothing is gained; the problem is to indicate a precise characteristic of continuity that can serve as the basis for valid deductions."

Dedekind's coup is to locate this "precise characteristic" not in *L*'s density or cohesion but rather in an obverse property, *severability*, which in turn is a consequence of the Number Line's being ordered and successive, i.e. of every point on the N.L. being to the right of all smaller-numbered points and to the left of all larger-numbered points. This means that at any particular point, we can as it were cut[11] the Number Line into two parts, two mutually exclusive infinite sets,[12] *A* on the left and *B* on the right, where every rational number in *B* is greater than every rational number in *A*. Here "Continuity and Irrational Numbers" (which is, as mentioned, unusually chatty for a technical math paper) has a ¶-long aside where Dedekind sort of kicks at the ground and pretends to apologize for how lame and obvious the *schnitt* thing must seem, e.g.: "[T]he majority of my readers will be very much disappointed in learning that by this commonplace remark the secret of continuity is to be revealed." The remark he's

[11] Dedekind's own verb here is *geschnitten* (*n.* = *schnitt*), which apparently can connote everything from slicing to cleaving—a very concrete, physical word, and rather more fun to say than 'cut'.

[12] **IYI** The translation of "Continuity and Irrational Numbers" uses 'classes,' which was math's original term for sets. Cantor, Russell, et al. tend to say 'class'—although in truth Cantor also sometimes uses words that translate as 'manifold' or 'aggregate'. Command Decision: From here on out we're just going to use 'set'.

referring to is the logical converse of the above severability-statement, that is:

> If all points of the straight line fall into two classes such that every point of the first class lies to the left of every point of the second class, then there exists one and only one point which produces this division of all points into two classes, this severing of the straight line into two portions.

What this means is that by defining the members and bounds of sets *A* and *B*, you can define the value of the point at which we cut *L* into *A* and *B*. And, as you'll recall from §2c, defining a point is defining a number.

Except, since the Line in question is the Number Line and maps only the rational numbers, it's fair to ask just how a *schnitt* is going to help define irrational numbers, which numbers are of course the true "secret of continuity". That every rational number will correspond to a *schnitt* but not every *schnitt* will correspond to a rational number might seem like just a restatement of the assertion that we can't define irrationals in terms of rationals. The answer is that Dedekind can literally *build* the definition of each irrational number out of the characteristics of the two sets it divides the Line into. Here's how the method works.

Consider a *schnitt* on the N.L. dividing the whole ∞ of rational numbers into two sets *A* and *B* such that all members *b* of *B* are greater than all members *a* of *A*. More specifically, consider whether *A* and *B* here can have largest/smallest members.[13] Depending on where and how the *schnitt* is

[13] **IYI** If you're flipping back, notice how close this is to the Cantorian stuff about infinite sets' *orders* in §7c ff.

defined, there are only three possibilities, of which only one can be true. Possibility 1 = Set *A* has a largest member a' (as in e.g. if the cut's *A* contains all rational numbers ≤ 2 and *B* contains all rational numbers >2). Possibility 2 = Set *B* has a smallest member b' (as in e.g. if the cut's *A* contains all rational numbers < 2). Possibility 3 = There is neither a largest member of *A* nor a smallest member of *B*.*

***MINI-INTERPOLATION**

Re these being the only three options, you can probably already see that there's no way set *B* is ever going to have a largest member, because *B* comprises everything from the *schnitt* to ∞. And Dedekind's *L* is the real N.L. and also includes all the rationals stretching leftward from 0, so *A* encompasses everything from $-\infty$ to the *schnitt* and can't have a smallest member. In case you're wondering, though, why there can't be a Possibility 4 in which there's both a largest member a' of *A and* a smallest b' in *B*, there's an easy way to prove this isn't possible. It's a reductio proof, so we posit that there is both a largest a' and a smallest b'. But this means that there will exist a certain rational number, equal to $\frac{a' + b'}{2}$, that is both larger than a' and smaller than b', and thus can be a member of neither *A* nor *B*. But *A* and *B* have been defined as together containing *all* the rational numbers. So Possibility 4 is contradictory.

But so then why isn't Possibility 3 contradictory in the same way?

DRAMATIC END OF MINI-INTERP.

Stated informally, Dedekind's example for Possibility 3 is a *schnitt* whose set *A* contains all the negative rational numbers and all positive rationals x such that $x^2 < 2$, and whose set *B* contains all the positive rationals x such that $x^2 > 2$. If it can be proved that no rational number corresponds to this *schnitt*, we will have defined a certain irrational number, in this case the millennially incommensurable $\sqrt{2}$.[14]

We've already seen, in §2c, a proof that $\sqrt{2}$ isn't rational[15]; we could rest on that laurel. But Dedekind doesn't, and provides his own specific proof that the *schnitt* in Possibility 3's example corresponds to no rational number. Which here it is. We might all want to breathe deeply for a moment and get very relaxed and attentive. Dedekind's is a reductio proof and so starts by assuming that there is indeed a rational x corresponding to Poss. 3's *schnitt*. If this x exists, then by the definition of set A, either x is A's largest member or else it's larger than any member of A (meaning it's in B). Either way, any number that's *larger* than x (let's call such a number x^+) is, by definition, absolutely going to be in set B, which means that $(x^+)^2$ must be >2. If, then, for any suitable x, we can produce an x^+ greater than x where $(x^+)^2 < 2$, the initial assumption that x is rational will be contradicted.

So accept the above specs on x and x^+, and define some positive number p as equal to $(2 - x^2)$, and define x^+ as

[14] **IYI** Dedekind's real example in "C. and I.N." is more abstract and uses the fact that if D is a positive integer that is not the square of any integer, then there exists a positive integer λ such that $\lambda^2 < D < (\lambda + 1)^2$, which happens now to be a basic theorem of number theory. Dedekind wants a totally abstract, general proof because his real goal is quote "to show that there exist infinitely many cuts not produced by rational numbers." His own proof is too hairy and number-theoretic for our purposes, though, so we'll just use $\sqrt{2}$ and ask you to trust that this result can be extended to cover all irrationals (which it can).

[15] **IYI** See or recall the incommensurability-of-$\frac{D}{S}$ demo on pp. 77–78.

equal to $(x + \frac{p}{4})$. That last def. may look a bit strange, but you can easily verify that given the original specs and value of p, $(x + \frac{p}{4})$ will be greater than x, so the crucial $x^+ > x$ thing is preserved. Now the rest of the proof is just good old tedious 8th-grade math:

(1) $(x^+)^2 = (x + \frac{p}{4})^2$

(2) $(x + \frac{p}{4})^2 = x^2 + \frac{xp}{2} + \frac{p^2}{16}$. Since x is by definition greater than 1, $x^2 > x$, so $(\frac{x^2p}{2}) > (\frac{xp}{2})$, so

(3) $(x^2 + \frac{xp}{2} + \frac{p^2}{16}) < (x^2 + \frac{x^2p}{2} + \frac{p^2}{16})$. Multiplying this larger quantity through to make 16 the common denominator, you get

(4) $(\frac{16x^2 + 8x^2p + p^2}{16})$. Since, by definition, $p = (2 - x^2)$, then clearly $x^2 = (2 - p)$; so by substitution the quantity in (4) becomes

(5) $(\frac{16(2 - p) + 8(2 - p)p + p^2}{16})$, which after distributing the constants becomes

(6) $(\frac{32 - 16p + (16 - 8p)p + p^2}{16})$, which after distributing the 2nd term's p then becomes

(7) $(\frac{32 - 16p + 16p - 8p^2 + p^2}{16})$, which quite obviously reduces to

(8) $(\frac{32 - 7p^2}{16})$, which is the same as

(9) $(2 - \frac{7}{16}p^2)$, which quantity, given that p is by definition > 0, is always going to be < 2. So by steps (1)–(9) (with particular highlights on (1), (3), and (9)), we have established that

(10) $(x^+)^2 < (2 - \frac{7}{16}p^2) < 2$, which by the basic law of transitivity means that

(11) $(x^+)^2 < 2$, which is precisely what we needed in order to contradict the initial assumption that the *schnitt* corresponded to a rational x. Meaning it doesn't correspond to a rational x. Q.E.D.

Dedekind, right after the proof: "In this property that not all cuts are produced by rational numbers consists the incompleteness or discontinuity of the domain of all rational numbers." But this isn't the half of it. The *schnitt* device allows irrational numbers to be defined wholly in terms of the rationals, which is the only way to engineer a 100% rigorous, deductive theory of real numbers. The definition, in English, is that an irrational number is the value of a point at which a *schnitt* divides the N.L. into a bilaterally exhaustive set *A* and set *B* that have no largest or smallest member respectively.[16] It's this definition that creates the Continuum—i.e., the set of all real numbers—and transforms the Number Line into the Real Line.[17] What's especially cool and ingenious is

[16] **IYI** Hence Dr. Goris's own classroom def. of a surd as a '*schnitt* sandwich,' which obviously went over big with adolescents.

[17] Anent a tossoff on p. 201: That the points on the R.L. can be put into one-to-one correspondence with the set of all real numbers is what's now known as the *Cantor-Dedekind Axiom.*

that Dedekind's technique uses just what had made surds so mysterious—their correspondence to unnamable points on the N.L.—as part of their rigorous def.

§6d. Of course, Dedekind's theory also presupposes the existence of actually-infinite sets. More than presupposes—via the *schnitt* device, the formal definition of a real number becomes 'A certain pair of infinite sets with certain particular characteristics'. There are a number of potential weirdnesses to notice at this point. First, given math's longstanding and much-referenced allergy to actual ∞s, Dedekind's theory might well be seen as simply trading one indefinable-type quantity for another—i.e. as invoking, in order to ground and define surds, the occult idea of not one but two unimaginably huge and yet precisely ordered sets, each one somehow infinite and yet limited in very specific ways. Which might strike you as just too much like Zeno all over again. If so, hold that thought.

Second weirdness: If you're unusually attentive and unborable, you might already have remarked a striking resemblance between Dedekind's theory and Eudoxus of Cnidos's geometric-commensurability thing back in §2d. With respect to which please recall or review §2d and see that the *schnitt* concept now helps make it clearer how Eudoxus's definition of 'ratio' works to designate irrational numbers: the number expressed by the ratio $\frac{p}{q}$ is irrational (that is, p and q are incommensurable) just when, for any rational number $\frac{a}{b}$, the disjunction $(ap < bq)$ or $(ap > bq)$ is true,[18] meaning when $ap \neq bq$. There's no imputation here that Dedekind was

[18] **IYI** Logicwise, a disjunction is true if at least one of its terms is true.

ripping Eudoxus off or necessarily even knew who he was. See for example the Preface to "The Nature and Meaning of Numbers," in which Dedekind cites the *Elements*'s Eudoxian Def. 5 without any evident awareness of where Euclid got it. This citation points up the big difference between Dedekind and Eudoxus, as well as between Greek math and modern analysis:

> [I]f one regards the irrational number as the ratio of two measurable quantities,[19] then is this [=Dedekind's own] manner of determining it already set forth in the clearest[20] possible way in the celebrated definition which Euclid gives of the equality of two ratios (*Elements*, V, 5).

The difference lies in the opening "If . . . measurable quantities" clause. Eudoxus and Euclid were (once again) geometers, and for them the problem of irrationals concerned geometric magnitudes like lines/areas/volumes. Whereas Dedekind's overall project (like, again, Weierstrass's own) is to get away from the geometry and ground analysis wholly in arithmetic.[21]

[19] This is D.'s term for geometric magnitudes.

[20] ulp

[21] **IYI** A related difference between Eudoxus and Dedekind is the way they conceive their respective theories' infinite sets. Recollect from later on in §2d that Eudoxus's Exhaustion Property involves infinite sets in the sense of infinite sequences of remainders-from-subtractions, except of course here the subtractions are from geometric magnitudes and the ∞ is only potential—as in, recall from p. 85, *Elements* X, Prop. 1's '$\lim_{n \to \infty} p(1 - r)^n = 0$,' which is just the way pre-Dedekind analysis would have treated Exhaustion's infinite remainders.

Mathematically and metaphysically, Dedekind's view is the exact opposite of Eudoxus's. Dedekind regards his own theory's geometric line/point thing as only theoretical and illustrative: it doesn't matter whether you could

This is why Dedekind says over and over again that the N.L. and geometric points in his *schnitt* theory are For Entertainment Purposes Only. His aforementioned Preface actually has one of the most stirring statements ever on the aesthetics of Arithmetization, viz.: "All the more beautiful it appears to me that without any notion of measurable quantities and simply by a finite system of simple thought-steps man can advance to the creation of the pure continuous number-domain."

On the other hand, whether it's OK to say that deploying actually-infinite sets in a mathematical definition involves only "a finite system of thought-steps" is a fair question, which returns us to the issue we tabled two ¶s back. If it so happens that you object to the use of infinite sets in a rigorous definition because you feel that such actual-∞-type sets are mathematically unreal/illicit, then you are of course an Aristotelian-slash-Gaussian and will count as your #1 advocate Prof. L. Kronecker (1823–1891), who as mentioned was G. Cantor's one-time mentor and later his arch nemesis and the person some historians think more or less singlehandedly drove him insane, and who (=Kronecker) was pretty much math's first *Intuitionist*, and believed that only integers were mathematically real because only they were 'clear to the intuition,' which meant that decimals, irrationals, and quite certainly infinite sets were all mathematical unicorns. Kronecker

---

ever really construct a complete Number Line or measure the exact lengths needed to isolate the point $\sqrt{2}$. What do matter—and are for Dedekind *actual*, since after all "Numbers are free creations of the human mind" and mathematical existence is "an immediate result from the laws of thought"—are the infinite sets A of all $x^2 < 2$ and B of all $x^2 > 2$. It is these ∞s, and not any geometric figures or quantities, that are fundamental to Dedekind's theory.

is often nutshelled in math history by his apothegmatic "The integers alone are created by God; all else is the work of Man," the same way that d'Alembert gets encapsulated by "Keep moving and faith will come" and Archimedes by "*Eureka!*"[22] There is more than you probably want to hear

[22] **IYI** The other big apocryphum about Kronecker concerns his reaction to F. Lindemann's proving in 1882 that π was a transcendental irrational (which proof finally put a stake through the heart of the Greeks' old Squaring the Circle problem). The story has Kronecker coming unbidden up to Lindemann at a conference and saying, very earnestly, "What use is your lovely proof about π? Why waste your time with problems like these, since irrationals don't even exist?" Lindemann's reaction is not recorded.

Quick profile of L. Kronecker *l'homme.* B.-D. dates already given. One of very few top mathematicians who was also a great businessman, Kronecker makes such a fortune in banking that at 30 he can retire and devote his life to pure math. Becomes a prof at U. Berlin, the Harvard of Germany, where he'd also been a star student. Researchwise, he's mostly an algebrist, his specialty being algebraic number-fields, which are a long story—likewise his most famous discovery, the Kronecker Delta Function, which in some ways anticipates the binary math of modern digitation. A serious gymnast and mountain-climber, Kronecker is no more than 5′0″ and lithe and muscular and always immaculately coiffed and dressed and accessorized. No kidding about the 5′0″. Courtly of manner, apparently, even though he's a total piranha in math- and academic politics. Very active and well-connected in the whole math community—the sort of colleague you do not want as an enemy because he's on every committee and editorial board there is. Main ally: the number-theorist E. E. Kummer. Main pre-Cantor enemy: the prenominately tall, rumpled, phlegmatic K. Weierstrass. Descriptions of Kronecker-v.-Weierstrass debates often feature the image of a rabid Chihuahua going after a Great Dane. Reason for antipathy: (1) Weierstrass's specialty is continuous functions, which Kronecker thinks are delusory and evil; (2) Kronecker believes that Weierstrass's Arithmetization of Analysis program doesn't go near far enough—see main text below. Kronecker's big dream: basing all of analysis

about L. Kronecker and Intuitionism a few §s down; for now it's enough to know that just as Weierstrass, Dedekind, et al. want to eliminate geometry from analysis and base everything on the real-number system, Kronecker goes even further and wants to base analysis only on integers and on rationals expressed as ratios of integers.

So now back to the specific can't-use-infinite-sets-in-a-definition objection to Dedekind's theory of real numbers. There are at least two nontrivial ways to respond to this. The first is to say (*pace* FN 21) that Dedekind's theory does not really require us to handle infinite sets in §1's special sense of 'handle'. Strictly speaking, the *schnitt* technique doesn't presuppose actual ∞s any more than '$\lim_{n\to\infty} (1 - \frac{1}{2^n})$' does. Which is to say that the infinite sets A and B in Dedekind's theory are entirely abstract, hypothetical: we don't have to count them or picture them or even think about them beyond knowing that $B > A$ and deciding the largest a'–v.–smallest b'–v.–neither issue.

If you are unmoved by this response and still feel that all Dedekind's supposedly rigorous def. is doing is trading irrational numbers' mathematical fuzziness for that of infinite sets, then it is appropriate to point out that right at the time "Continuity and Irrational Numbers" comes out, G. F. L. P. Cantor is beginning to publish work that defuzzifies just the sorts of actually-infinite sets Dedekind is positing. As was

---

on the integers. K.'s ultraconservative math-ontology is now regarded as the forerunner of *Intuitionism* and *Constructivism*, though it had no other adherents until Poincaré and Brouwer—all of which gets unpacked at length in §7.

foreshadowed in §§ 1 and 4, Cantor and Dedekind's near-simultaneous appearance in math is more or less the Newton + Leibniz thing all over again, a sure sign that the Time Was Right for ∞-type sets. Just as striking is the Escherian way the two men's work dovetails.[23] Cantor is able to define and ground the concepts of 'infinite set' and 'transfinite number,' and to establish rigorous techniques for combining and comparing different types of ∞s, which is just where Dedekind's def. of irrationals needs shoring up. Pro quo, the *schnitt* technique demonstrates that actually-infinite sets can have real utility in analysis. That, in other words, as sensuously and cognitively abstract as they must remain, ∞s can nevertheless function in math as *practical* abstractions rather than as just weird paradoxical flights of fancy.[24]

Also almost eerie is the timing. Dedekind first learns of G. Cantor's existence in March, 1872, when he reads the latter's "*Über die Ausdehnung eines Satzes aus der Theorie der trigonometrischen Reihen*"[25] in a big journal as he's putting the finishing touches on "Continuity and Irrational Numbers," into the last draft of which he sticks a Cantor-citation and a "hearty thanks" to "this ingenious author" whose paper's own theory of irrationals* ". . . agrees, aside from the form of presentation, with what I designate as the essence of

[23] **IYI** Dr. Goris liked to say that Weierstrass, Dedekind, and Cantor composed their own special convergent series, with each in turn supplying just what the others needed to make their advances viable.

[24] There are, it goes without saying, late-nineteenth-century mathematicians to whom none of this is convincing. Kronecker is only one of them.

[25] = roughly "On the Extension of a Proposition from the Theory of Trigonometric Series," which is C.'s first important article and will occupy much of the next two §s.

continuity." Then later that same year they meet, by accident, at some vacation retreat in Switzerland. They literally bump into each other. Cantor is a *privatdozent* at the University of Halle, Dedekind teaching high school in Brunswick.[26] They hit it off, and start exchanging letters—and a lot of Cantor's most significant results get worked out in these letters. But the two are never true collaborators, and there's apparently a big falling out in the 1880s when Cantor finagles Dedekind a professorship at Halle and R. D. turns it down (although they must have eventually made up if they're having day-long lunches together in 1899). Again, most of this sort of personal stuff we're skipping.

§6e. *SEMI-IYI INTERPOLATION

G. Cantor's theory of irrationals, which as mentioned is mostly in '72's "*Über . . . Reihen*," is both technically complex and ultimately less significant than the larger set-theoretic work it's part of,[27] and depending on your overall $\frac{\text{interest}}{\text{fatigue}}$ ratio you may wish to do no more than skim the following gloss, which occupies a tricky rhetorical niche and has been classed **SEMI-IYI**.

Cantor is less concerned with defining irrationals per se than with developing a technique by which he can define all real numbers, rational and irrational, in the same way. It's

[26] **IYI** Gorisian factoid: Brunswick is best known for its inexplicably popular Braunschweiger sandwich spread, which is a little like pâté that's been allowed to solidify in the toe of an old gym-sneaker, and which during our class's Dedekind-Cantor unit everyone was invited (for Extra Credit) to try a little of on a Wheat Thin.

[27] More to the point for us is that Cantor's theory of irrationals just isn't as good as Dedekind's. The latter is simpler and more elegant and (rather ironically) makes better use of ∞-type sets.

actually Cantor who first introduces to math the idea of a set of all real numbers comprising both the rationals and the surds (which Dedekind, for complicated reasons, objects to). Cantor's own theory obviously depends on infinite sets, too; but for Cantor at this point these are more like infinite sets of infinite sequences of rational numbers. Hence part of his theory's hairiness. Another part is that Cantor wants to use the Weierstrassian idea of irrationals-as-limits without the circularity problem mentioned in §6a, which leads him to use convergent sequences rather than series.

More specifically, in order to avoid Weierstrass's def.'s circularity, Cantor exploits the facts that (1) all real numbers can be represented by infinite decimals (these rational decimals always either infinitely repeating their basic period (as in §6c FN 10) or else ending in an ∞ of 0s or 9s (which, retrieving the lagan from §2c FN 35, are equivalent, as in the whole 0.999 . . . = 1.000 . . . thing)), and that (2) infinite decimals function as limits of *decimal fractions.* (You learned decimal fractions in 4th or 5th grade; they're the way we teach kids to make sense of decimals in terms of fractions, as in for instance $0.15 = \frac{0}{1} + \frac{1}{10} + \frac{5}{100}$. Here, then, is another place where high-end analysis turns out to underlie childhood arithmetic: the general rule is that any infinite decimal can be represented as the convergent infinite series $a_0 10^0 + a_1 10^{-1} + a_2 10^{-2} + a_3 10^{-3} + \cdots + a_n 10^{-n} + \cdots$, which series sums/converges to the original decimal—i.e., the decimal is the series' limit.[28]) Then, through a clever bit of

[28] **IYI** You'll likely notice some marked similarities between the stuff on decimal fractions and §5e(1)'s thing about approximating rational numbers via convergent power series of other rationals. Rhetoricwise, let's

math-semantics, Cantor can bag the whole decimals thing by observing, in the spirit of §5d, that any sequence[29] of rational numbers' *converging* is the same as its being *representable by* an infinite decimal, which decimal is thus mathematically *defined by* the sequence.

(**IYI** If the above ¶ seems shifty or convoluted, we can reduce the argument to a simple syllogism: 'Since (1) all numbers are definable by decimals and (2) all decimals are definable by sequences, (3) all numbers are definable by sequences,' which happens to be 100% valid.)

Given all this, Cantor's basic idea is that an infinite sequence $a_0, a_1, a_2, \ldots, a_n, \ldots$ of rational numbers defines a real number if the sequence converges such that $\lim_{n\to\infty}(a_{n+m} - a_n) = 0$ for any arbitrary m[30]—that is, if the difference between any two successive terms tends to 0[31] as you get farther and

---

concede one more time that if we were after technical rigor rather than general appreciation, all these sorts of connections would be fully traced out/discussed, though of course then this whole booklet would be much longer and harder and the readerly-background-and-patience bar set a great deal higher. So it's all a continuous series of tradeoffs.

[29] **IYI** Surely it's unnecessary to stress once again that a series is just a particular type of sequence.

[30] **IYI** If you've noticed how similar this is to §5a's Cauchy Convergence Condition (q.v. E.G. II, —*Differential Equations* (b)), you can understand why Cantor doesn't say the *a*-sequence defines a real number *if and only if* it converges in this way, but merely *if* it converges.

[31] Naturally this is just shorthand. In the real "*Über . . . Reihen,*" Cantor follows Weierstrass's program and eschews 'tends to'/'approaches' in favor of the good old little epsilon; so technically the sequence's limit is defined by the rule that for any given ε, no more than a finite number of the sequence's successive terms can fail to differ from each other by a value m such that $m < \varepsilon$.

farther out in the sequence (which of course is just the way decimals really work: by the nth decimal place, successive values can differ by at most 10^{-n}). Cantor calls sequences that behave this way *fundamental sequences*, and what's known as Cantor's Theory of Real Numbers is that each real number is defined by at least one fundamental sequence.

A couple potential objections at this point. If it strikes you as maybe a bit circular or question-begging for Cantor to define 'real number' as that-which-is-defined-by-a-fundamental-sequence—rather like defining 'dog' as that-which-is-defined-by-the-definition-of-'dog'—then we have to get clear on just what 'define' means w/r/t Cantor's theory. It pretty much means 'is' or 'equals'. That is, what enables the theory to avoid circularity is that the relevant fundamental sequence *is* the real number, the same way that 0.15 *is* Lim $(0(10^0), 1(10^{-1}), 5(10^{-2}))$ and a trig function *is* its convergent-series expansion. On the other hand, since the real numbers include both the irrationals and the rationals, you might be wondering whether and how Cantor's fundamental sequences of rationals can also define rational numbers, or whether the idea even makes sense. The answer is that it does and they can: Cantor's stipulation is that when a fundamental sequence $a_0, \ldots, a_n, \ldots$ has each of its terms after a_n equal to either 0 or a, the sequence defines (= *is*) the rational number a.[32]

In order to make his theory truly viable, Cantor still has to show how to prove arithmetical properties and perform basic operations with his fundamental sequences and the real

[32] **IYI** Example of a fundamental sequence where $a_{n+1} = 0$: 0.1500000 . . .; example where $a_{n+1} = a$: 0.66666 . . ..

numbers they define. What follow are a couple examples of his paper's demonstrations, with our relevant real numbers here being x and y:

(A) Since it turns out that you can define the same real x via more than one fundamental sequence,[33] Cantor's rule is that two fundamental sequences $a_0, a_1, a_2, \ldots$ and $b_0, b_1, b_2, \ldots$ define the same x if and only if $|a_n - b_n|$ approaches[34] 0 as n tends to ∞.

(B) To prove basic arithmetical operations, let's say a_n and b_n are fundamental sequences defining x and y respectively. Cantor proves (in an orgy of high-tech symbolism that we're omitting) that $(a_n \pm b_n)$ and $(a_n \times b_n)$ are also fundamental sequences, thus defining the real numbers $x \pm y$ and $x \times y$. In his proof that $\frac{b_n}{a_n}$ is a fundamental sequence defining $\frac{y}{x}$, the only restriction is that x obviously can't be 0.

As a sort of wrap-up, you might have noticed one other potential objection. A rather nasty one. And perhaps the single most impressive thing about Cantor's theory of reals-via-fundamental-sequences—and what makes Dedekind call him ingenious—is the way Cantor avoids this lethal VIR-type objection that at first his theory looks vulnerable to. For if fundamental sequences of rational numbers define real numbers, what about fundamental sequences of *real* (= both rational and irrational) numbers? You can easily construct a

[33] infinitely many, as a matter of fact, though the proof's too involved to get into.

[34] Shorthand again. Same with the next verb phrase.

real-number sequence that converges according to Cantor's specs—many kinds of trig series already do. Do we need to create a whole new class of numbers to function as the limits of these real-number sequences? If we do, then we'll need still another class to serve as the limits of fundamental sequences of those new numbers, and then yet another . . . and off we go; it'll be Aristotle's Third Man all over again. Except Cantor heads this off by proving[35] the following theorem: If r_n is a sequence of real numbers such that $\lim_{n\to\infty} (r_{n+m} - r_n) = 0$ for an arbitrary m—that is, if r_n is a bona fide fundamental sequence of reals—then there is some unique *real* number r, defined by a fundamental sequence a_n of rational a's, such that $\lim_{n\to\infty} r_n = r$. In other words, Cantor is able to show that real numbers *themselves* can serve as the limits of fundamental sequences of reals, meaning his system of definitions is self-enclosed and VIR-proof.

§6f. As was vaguely foreshadowed in §6d, Dedekind's and Cantor's theories run afoul of yet another Kroneckerian doctrine, often known as *Constructivism*, which will become a big part of Intuitionism and of the controversies over math's philosophical foundations touched off by set theory.[36] This all gets

[35] again, offstage—it would take several pages and a whole other EMERGENCY GLOSSARY to unpack it. We'll be spending more than enough time on Cantor's formal ∞-proofs in §7 as it is.

[36] **IYI** Some math-historians use 'Constructivism' to mean also Intuitionism, or vice versa, or sometimes *Operationalism* to mean both C.ism and I.ism—plus there's *Conventionalism*, which is both similar and somewhat different—and the whole thing can just get extremely convolved and unpleasant. From here on, we're going to classify the various schools and

very heavy and complicated, but it's important. Here are the basic principles of Constructivism as practiced by Kronecker and codified by J. H. Poincaré and L. E. J. Brouwer and other major figures in Intuitionism: (1) Any mathematical statement or theorem that is more complicated or abstract than plain old integer-style arithmetic must be explicitly derived (i.e., 'constructed') from integer arithmetic via a finite number of purely deductive steps. (2) The only valid proofs in math are constructive ones, with the adjective here meaning that the proof provides a method for finding (i.e., 'constructing') whatever mathematical entities it's concerned with.[37]

---

ideologies in as nongrotesquely simple a way as possible. (Please N.B. also that we're going to be discussing no more of the whole Intuitionist controversy than is directly mission-relevant. For readers especially interested in the metamathematical debate that starts with Kronecker and ends with Gödel's foundational demolitions in the 1930s, we're pleased to recommend P. Mancuso, Ed.'s *From Brouwer to Hilbert*, S. C. Kleene's *Introduction to Metamathematics*, and/or H. Weyl's *Philosophy of Mathematics and Natural Science*, all of which are in the Bibliography.)

[37] There's an interesting English-language coincidence in the Kroneckeroid idea of constructive proof. As well as literal connotations like 'involving actual construction,' the word 'constructive' for us can mean 'not destructive'. As in good rather than bad, building up rather than tearing down. The main kind of destructive proof happens to be the reductio ad absurdum; and sure enough, Constructivism does not consider the reductio a valid proof-procedure. The real reason, though, is that the reductio depends logically on the Law of the Excluded Middle, by which (as you'll recall from §1c) every math-type proposition is, formally speaking, either true or if not true then false. Constructivists (esp. the extremely eccentric and humorless L. Brouwer) reject LEM as a formal axiom, mainly because LEM can't be constructively proven; that is, there is no stated decision procedure by which you can verify, for *any* proposition P, whether P is

Respecting the metaphysics of math, Constructivism is therefore directly opposed to Platonism: except for maybe the integers, mathematical truths do not exist apart from human minds. In fact, as far as Kronecker et seq. are concerned, to say that a certain mathematical entity 'exists' is literally to say that it can be constructed w/ pencil and paper by real human beings within mortal timespans.

You can see, then, that Constructivists are going to have a serious problem with theorems and proofs that involve ∞, infinite sets, infinite sequences, etc.—particularly when these infinite quantities are explicitly presented as actual. With respect to Dedekind's *schnitts*, for instance, the Constructivists' shorts are obviously going to be in a knot from the outset. Not only do irrationals not really exist, and not only does Dedekind use reductio to prove that some *schnitts* don't correspond to rational numbers. There's also the whole problem of defining a number in terms of infinite sets of other numbers. For one thing, Dedekind doesn't ever specify the mathematical rules by which one derives the sets *A* and *B*.

---

true or false. Plus there are all sorts of important math hypotheses—such as e.g. Goldbach's Conjecture, the Irrationality of Euler's Constant, and Cantor's own upcoming Continuum Hypothesis—that either can't be proved yet or can be shown to be LEM-grade unprovable. Etc. etc. (Note, **IYI**, that however cranky or fundamentalist Constructivism might sound, it's not without influence and value. The movement's emphasis on decision procedures was important to the advent of mathematical logic and computer design, and its rejection of LEM is part of why Intuitionism is regarded as the forefather of multivalued logics, including the Fuzzy Logics so vital to today's A.I., genetics, nonlinear systems, etc. (**IY very I** Regarding that last item, readers with hypertrophic math backgrounds and a lot of time on their hands should see Klir and Yuan's *Fuzzy Sets and Fuzzy Logic: Theory and Applications*, whose specs are also in the Bibliography.))

He simply says *if* the Number Line can be divided into A and $B \ldots$, without giving any method or procedure by which one could actually construct these sets—sets that can't really be constructed or verified anyway, since they're infinite. And while we're at it, just what exactly is a 'set,' technically speaking; and what's the procedure for constructing one? And so on.

That last Constructivist compound-question (which admittedly Dedekind can't answer[38]) affords a prime instance of G. Cantor Jr.'s particular genius and of why he deserves the title Father of Set Theory. Recall §3c's mention of how Cantor took what had been regarded as a paradoxical, totally unhandlable feature of ∞—namely that an infinite set/class/aggregate can be put into a one-to-one correspondence with its own subset—and transformed it into the technical def. of *infinite set.* Watch how he does the same thing here, turning what appear to be devastating objections into rigorous criteria, by defining a *set* S as any aggregate or collection of discrete entities that satisfies two conditions: (1) S can be entertained by the mind *as* an aggregate, and (2) There is some stated rule or condition via which one can determine, for any entity x, whether or not x is a member of S.[39]

This definition doesn't just suddenly appear out of the blue, obviously. It now becomes appropriate to revisit §5d's stuff about trig-series convergence and representability, Riemann's Localization Theorem, etc., in order to see where Cantor's work on surds and sets really comes from.

[38] although in all fairness it's not really his area.

[39] **IYI** There's more detail/context on this def. coming up in §7a. For now, a good example of a *set* would be the set of all irrational numbers, especially since we just saw Dedekind spell out a very definite procedure for determining whether any given number is irrational or not.

§7a. This section even has epigraphs:

> "The modern theory of the infinite evolved in a contiguous way out of the mathematics that preceded it."
>
> —*S. Lavine*

> "But the uninitiated may wonder how it is possible to deal with a number which cannot be counted."
>
> —*B. Russell*

> "Buckle your lapbelts please everyone as we are about to undergo a steep ascent."
>
> —*E. R. Goris*

For reasons that by now will be familiar, a lot of what follows is going to be really fast. It's also somewhat hard at the start, but like a lot of pure math it gets easier the deeper we go. As mentioned already, G. F. L. P. Cantor does his graduate work at Berlin under Weierstrass and Kronecker; his earliest published articles are fairly standard-issue work in certain problems of number theory.[1] Ph.D. in hand, Cantor gets a low-rung job as a *privatdozent* (which seems to be a weird sort of freelance T.A.[2]) at U. Halle, and there meets one E. H. Heine (1821–1881), a specialist in applied analysis who'd done important work in Heat, particularly on the

[1] **IYI** More specifically in indeterminate equations and ternary forms, both of which are algebraic number-theory topics that Kronecker's interested in. (Have we mentioned that Kronecker was Cantor's dissertation advisor? and that it's SOP for young mathematicians to work on mentors' problems—q.v. Cantor also taking over Heine's proof just below?)

[2] **IYI** The German academic system of the 1800s is pretty much unparsable.

Potential Equation.[3] Anyway, as of c. 1870 Heine is part of the whole big group of mathematicians working on Fourier Series and the issues raised by Riemann's "On the Representability . . .," and apparently he (=Heine) gets Cantor interested in what had come to be known as the Uniqueness problem: If any given $f(x)$ can be represented by a trigonometric series, is that representation Unique, i.e. is there only one trig series that can do it? Heine himself had been able to prove Uniqueness only on the condition that the relevant trig series was uniformly convergent.[4] Which clearly wasn't good enough, since there are plenty of trig series and even Fourier Series that aren't uniformly convergent.

In 1872, Cantor's "On the Extension of a Proposition from the Theory of Trigonometric Series"[5] defines and proves a far more general Uniqueness Theorem that not only doesn't require uniform convergence but permits exceptions to convergence at an infinity of points, provided that these exceptional points[6] are distributed in a certain specific way.

[3] **IYI** = a particular kind of partial differential equation that you might remember from sophomore math, probably in the context of Green's Theorem.

[4] which, you might recall from E.G.II's thing on —*Uniform Convergence* & Associated Arcana, means that the $f(x)$ the series represents (=sums to) must be continuous. (**IYI** N.B.: Heine's real proof requires that the series and function be '*almost everywhere*' uniformly convergent and continuous, which involves distinctions so fine that we can ignore them without distortion.)

[5] **IYI** i.e., the same paper in which he lays out his theory of real numbers in §6e (which paper's title will probably now make more sense).

[6] **IYI** Q.v. E.G.II's —*Uniform Convergence* & Associated Arcana item (d) for exceptional points, which again please recall can also be called 'discontinuities'. (N.B.: Some math classes also use *singularity* to mean exceptional point, which is both confusing and intriguing since the term also refers to Black Holes, which in a sense is what discontinuities are.)

As we'll see, this precise distribution's rather complicated—as is the '72 Theorem itself, which Cantor has actually developed over several prior papers and published appendices to same, as his criteria for Uniqueness gradually evolve from requiring the given trig series to converge for all values of x, to allowing a finite number of exceptional points, to the Uniqueness Theorem's final, 100% general form. What's interesting is that it's Professor L. Kronecker who helps Cantor refine and simplify his proof at several early points. At the same time, Cantor's approach is deeply informed by K. Weierstrass's work on continuity and convergence, e.g. the observation that for a general trig series of the form $f(x) = \frac{a_0}{2} + \Sigma(a_n \sin nx + b_n \cos nx)$ to be integrable term-by-term (which was the way Heine and everyone else had tried to attack the Uniqueness problem), it has to be uniformly convergent. Historians have noted that Cantor-Kronecker relations cooled and L. K.'s letters got more and more critical as Cantor's refinements of the Uniqueness Theorem began to allow infinitely many points at which exceptions to either the representation of the given $f(x)$ or the convergence of the series[7] could be allowed. In each successive version of the proof, Kronecker was basically watching Cantor move from his own algebraic, Constructivist position to a more Weierstrassian, function-theoretic one. The apostasy was complete when the final 1872 paper came out and its whole first part was devoted to the theory of irrationals/reals detailed in §6e.

Understanding just why it required a coherent theory of irrationals is as good a way as any to see how Cantor's work

[7] It is important throughout this section to remember that as far as we're concerned these are the same thing.

on the Uniqueness Theorem led him into studying infinite sets per se. The discussion, which gets a bit hairy, requires that you recollect the Bolzano-Weierstrass Theorem's rule that every bounded infinite set of points contains at least one limit point—and so of course what a limit point is.[8] Plus you need to keep in mind that the Uniqueness proof's central concepts are all about the Real Line (i.e., when terms like 'point set' or 'exceptional point' or 'limit point' are used, they are really referring to geometric points that correspond to numbers[9]). Please note also that the 'bounded infinite sets of points' under consideration in both the B.W.T. and Cantor's proof are actually sequences, which are also the easiest entities w/r/t which to understand limit points—like for instance the infinite set of N.L. points $\{0, \frac{1}{4}, \frac{3}{8}, \frac{7}{16}, \frac{15}{32}, \frac{31}{64}, \cdots\}$[10] has as its limit point the same $\frac{1}{2}$ that is the limit of the infinite sequence $0, \frac{1}{4}, \frac{3}{8}, \frac{7}{16}, \frac{15}{32}, \frac{31}{64}, \ldots$.

So here's how Cantor gets his increasingly general result. At first (1870), he needs a trig series that isn't uniformly convergent[11] but is convergent everywhere, meaning convergent for all values of *x*. In the next step (1871)[12], he is able to prove

[8] **IYI** q.v. §5e, p. 188's text and FN 72.

[9] C.f. Cantor, right at the start of "On the Extension . . . Series": "To every number there corresponds a definite point of the line, whose coordinate is equal to that number."

[10] Have we inserted yet anyplace that these strange human-profile brackets, '{ },' are what you put around things to show they compose a mathematical set?

[11] i.e., that is not integrable term by term such that the result is convergent.

[12] **IYI** Evidently here's where Kronecker was especially helpful.

that if two apparently distinct trig series converge to the same (arbitrary) function everywhere except a finite number of x-points, they are really the same series. We're going to skip this proof because what's really germane is the next one, which is the 1872 result where Cantor is able to allow an infinite number of exceptional points and still prove that the two representative trig series are ultimately identical.[13] How he does this is by introducing the concept of a *derived set*, whose def. is basically: If P is a *point set* (meaning just any set of real-number points, though what Cantor obviously has in mind is the infinite set of all exceptional points between the two trig series), then P's *derived set* P' is the set of all limit points of P. Or rather we should say that P' is the *first* derived set of P, because as long as the relevant point sets are infinite, the whole process is, in principle, endlessly iterable—P'' is the derived set of P', P''' is the derived set of P'', and so on, until after $(n - 1)$ iterations you'll have P^n as the nth derived set of P. In the gloss of the Uniqueness Theorem two ¶s back, what the "provided that these exceptional points are distributed in a certain specific way" criterion centers on is this nth derived set P^n, with the vital question being whether P^n is infinite or not.

Some of its real math involves extremely technical stuff about how limit points operate, but in essence here's how the U.T.'s exceptional-point-distribution thing works. Another deep breath would probably not be out of place.[14] Posit an

[13] And clearly, if any series-representation ends up being provably identical to the $f(x)$'s original representative series, then that original series is the function's Unique representation. This is basically how the Uniqueness Theorem works—it's not that there's only one series that can represent a given function, but rather that all series that do represent it are provably equivalent.

[14] **IYI** If the following couple ¶s seem brutal, please don't lose heart. The truth is that Cantor's route *to* the theory of ∞ is a lot harder, mathwise, than

infinite point set P such that, for some finite n, P's $(n-1)$th derived set $P^{(n-1)}$ is infinite while its nth derived set P^n is finite. Then, if two trigonometric series converge to the same $f(x)$ except at some or all of the points in P, they are the equivalent series; hence Uniqueness is proved. That was probably not blindingly vivid and clear. The part it's crucial to unpack is the requirement that P^n be finite. And to explain it, we have to introduce another distinction: any set P for which its derived set P^n is finite for some finite n is what Cantor calls a set of the *first species*; whereas, if P^n is not finite for any finite value of n, then P is a set of the *second species*. (This is why the derived-set process was described above as "in principle" endless—it's only the second-species sets that never produce finite derived sets and so allow a ∞ of iterations of the derived-set-of-derived-set thing.)

OK then. Here's why, in his proof of the U.T., Cantor needs the original infinite set of exceptional points P to be a first-species set, and thus P^n to be finite. It is because, via K. Weierstrass's Extreme Values Theorem, you can prove for sure that if any derived set P^n is finite, then at some further point $n+k$ the derived set $P^{(n+k)}$ is going to take its absolute minimum value m, which in this case will be 0, or no limit points at all. Meaning, in other words, that anytime in the whole P', P'', P''', . . . P^n, . . . progression you arrive at a derived set that's finite, you know that the whole iterative process is going to stop somewhere; you're eventually going to get to a $P^{(n+k)}$ with no members. And of course we know

---

the theory itself. All you really need to get is a rough sense of how the Uniqueness Theorem leads Cantor into transfinite math. The brutal part will be over quickly.

that the members of all these various P's and P''s are limit points; and we further know, from the Bolzano-Weierstrass Theorem, that any bounded infinite set has *at least one* limit point. If $P^{(n+k)}$ has no members, then the set of which it is the derived set has no limit point, and thus by the B.W.T. must itself be finite, and thus by the E.V.T. must at some point take its own minimum value of 0 members, at which same point the set of which *that* is the derived set becomes finite, and so on back down through the $P^{(n+k)}$s and P^ns and $P^{(n-1)}$s and P's . . . all of which means—to boil everything way, way down—that at some provable point the two representative trig series can be shown to collapse into a single series, which establishes Uniqueness.

You'll remember from §§ 5e and 6a, though, that in order to be 100% workable in all cases the Extreme Values Theorem requires a theory of real numbers, and that the Weierstrassian version of such a theory was (as Cantor himself showed) a dink. This is one reason why Cantor's 1872 proof requires its own theory of reals. The other, and related, reason is that in order to use the more general Bolzano-Weierstrass Theorem to construct his theory of derived sets and species, Cantor needs to reconcile the geometric properties of points on a line's continuum—meaning here concepts like 'limit point,' 'interval,' etc.—with the arithmetic continuum of real numbers, since the entities involved in analysis are of course really numbers and not points.[15]

[15] See for instance J. W. Dauben on the '72 proof: "Cantor stressed, however, that the numbers in these various [derived-set] domains remained entirely independent of this geometric identification, and the

If we observe that Cantor's derived sets resemble the general idea of a *subset*, and that the minimum value $P^{(n+k)} = 0$ is essentially the same as the definition of the *empty set*, it's possible to discern the seeds of what's now known as set theory[16] in Cantor's Uniqueness proof. Derived sets, the R.L./real-number continuum, and the Uniqueness Theorem are

---

isomorphism served, really, as an aid in thinking about the numbers themselves." You'll notice that this attitude is more or less identical to Dedekind's in "Continuity and Irrational Numbers".

[16] Preliminary tidying: We need to draw a distinction between two different kinds of set theory you might know about. What's called *point-set theory* involves sets whose members are numbers, spatial or R.L.-points, or various groups/systems of these. Point-set theory is today a big deal in, e.g., function theory and analytic topology. *Abstract set theory*, on the other hand, is so named because the nature and/or members of its sets isn't specified. Meaning it concerns sets of pretty much anything at all; it's totally general and nonspecific—hence the 'abstract'.

From here on out, 'set theory' will refer to abstract set theory unless otherwise specified.

What's complicated is that G. Cantor, whose real fame is as the author of abstract set theory, obviously started out in (and basically invented) the point-set kind. It's really not until the 1890s that he provides the definition of *set* that now characterizes abstract set theory—"A collection into a whole of definite, distinct objects of our intuition or thought, [which] are called the members of the set"—but all his significant results in the '70s and '80s on point sets also apply to abstract s.t. Plus finally please note, anent Cantor's above definition v. the gloss we saw in §6f, that his 'definite' means that for some set S and any object x, it's at least in principle possible to determine whether x is a member of S (if you've had much logic, you might recall that this feature of formal systems is called *Decidability*). Whereas his definition's 'distinct' means that for any two members x and y of S, $x \neq y$, which serves formally to distinguish a set from a sequence, since in sequences the same term can show up over and over. Tidying complete.

also the progenitors of Cantor's transfinite math, although in a rather complex bunch of ways. We've just seen that the Uniqueness Theorem's P^n requires n to be finite, i.e. that Cantor uses only finite iterations of the derived-set-of-derived-set process for his proof. Since there are already so many infinite sets floating around the proof, though (as in: the original P is infinite, and all the P's and P''s and so on up through $P^{(n-1)}$ can be infinite, and of course the relevant trig series are infinite series, not to mention that limit points involve an infinite number of intra-interval points, and that these intervals can themselves be infinitesimally small), it shouldn't be surprising that Cantor starts to consider more closely the characteristics of his derived sets under infinite iterations.

More specifically, Cantor starts to ask whether the infinity of an infinitely iterated second-species derived set P^∞ might differ from or somehow exceed the infinity of the finitely iterated first-species set $P^{(n-1)}$. More specifically still, he notices how closely those questions resemble one about the relative ∞s of the N.L.'s rational numbers versus the R.L.'s real numbers. This latter question concerns the issue first discussed all the way back in §2c—namely that while the rational numbers are both infinite and infinitely dense, they are not continuous (i.e., the Number Line is shot through with holes), whereas of course both Dedekind and Cantor have now proved the continuity of the set of all reals as schematized on the R.L. It thus appears natural to Cantor[17]—who for the U.T. has already developed techniques for examining both real v. rational

[17] **IYI** It goes without saying that this is all very condensed—it's not like there was some single epiphanic moment at his desk when these things occurred to him.

numbers and the properties of infinite sets—to ask whether the infinite set of all real numbers is somehow bigger than the infinite set of all rational numbers. Except what would 'bigger' here mean, and ditto for 'exceed' in the above P^{∞} v. $P^{(n-1)}$ question—that is, how can the comparative sizes of different ∞s be described and explained mathematically? At which point, in the immortal words of J. Gleason. . . .

ADMINISTRATIVE INTERPOLATION

There are now a couple procedural issues that need to be addressed. Whole scholarly tomes are devoted to Cantor's accomplishments.[18] You can take a two-term course on set theory under the departmental rubrics of Logic, Math, Philosophy, or Computer Science[19] and still have just scratched the veneer. Historically, Cantor's transfinite theories and proofs are spread over 20 years[20] and scores of different

[18] **IYI** First-rate scholarly books in English include A. Abian's *The Theory of Sets and Transfinite Arithmetic*, M. Hallett's *Cantorian Set Theory and Limitation of Size*, and J. W. Dauben's *Georg Cantor: His Mathematics and Philosophy of the Infinite*, all of which are listed in the Bibliography. The caveat, though, is that 'scholarly' here means pitilessly dense and technical. Dauben's book in particular requires such a strong pure-math background that it's hard to imagine any reader who's able to enjoy it wasting her time on the present booklet . . . which renders this whole FN self-nullifying in a sort of interesting way.

[19] **IYI** the last usually in connection with the extensional logic of G. Boole (1815–1864), whom it's a shame we're not going to talk about.

[20] **IYI** It's more nearly accurate to say that the bulk of his original work gets done 1874–84, with the subsequent decade's papers being mostly expansions and revisions of previous proofs, as well as responses to other mathematicians' objections.

papers, often with successive refinements and amendments to earlier stuff so that there is sometimes more than one version of the same proof. It's clearly going to be impossible here to unpack transfinite math all the way or to do real justice to its evolution in Cantor's publications.[21] On the other hand, there are certain recent pop books that give such shallow and reductive accounts of Cantor's proofs (accounts which are, as mentioned, usually subordinated to some larger Promethean narrative about G. Cantor's psych problems or supposed mystical affiliations) that the math is distorted and its beauty obscured; and we obviously don't want to do this, either.

So Command Decision: The compromise henceforth will be to sacrifice chronology and a certain developmental thoroughness for the sake of conceptual thoroughness and cohesion—that is, to present Cantor's concepts, theorems, and proofs in a way that highlights their connections to one another and to math per se. This will involve not only skipping around a bit, but often not telling you that we're skipping around, or that there are sometimes several different versions of a given proof and we're covering only the best one, or what the exact dates and English-v.-German titles of all the

[21] Please note also that as he was inventing the stuff Cantor did not proceed axiomatically; he based most of it on what he called "pictures," or rough concepts. Plus, despite the heavy symbolism, most of his actual arguments were in natural language—and not exactly crisp clear Russell-caliber language, either. The point being that Cantor's original work is quite a bit fuzzier and more complicated than the transfinite math we're going to discuss, much of which latter uses axiomatizations and codifications supplied by post-Cantor set theorists like E. Zermelo, A. A. Fraenkel, and T. Skolem (not that the names matter much yet in this §).

relevant articles are,[22] etc. It also entails a special set-theoretic EMERGENCY GLOSSARY III, though this E.G. will need to be administered gradually and *in situ* because some of the material is just too abstract to shove at you up front without context.

END A.I.

§7b. As should be evident, some very important ∞-related ideas come out of the proof of the general U.T. One concerns the relative sizes of the set of all rationals v. the set of all reals; another is whether the continuity of the Real Line is related somehow to the size/composition of the latter set. Yet another is the concept of a *transfinite number*, which Cantor derives from the same considerations that led him to distinguish first- from second-species infinite sets in the '72 proof.

To see how Cantor conceives and generates his transfinite numbers, we need first to make sure we're E.G.-grade clear on a few set-theory terms you probably first saw in elementary school.[23] To wit: Set A is a *subset* of set B if and only if there is no member of A that is not a member of B. The *union* of two sets A and B is the set of all members of A and all members of B, while the *intersection* of A and B is the set of just those members of A that are also members of B. Union and intersection are normally symbolized by '$\cup$' and '$\cap$,' respectively. Lastly, the good old *empty set*, whose usual symbol is '$\emptyset$,' is a

[22] **IYI** although obviously everything is ultimately findable in the Cantor publications listed in the Bibliography.

[23] **IYI** especially if you're the right age to have been subjected to the New Math.

set with no members—and be apprised that, by what looks at first like merely a quirk of the definition of 'subset,' any set whatsoever will always include Ø as a subset. End of first situated chunk of E.G.III. Plus here you have to remember the first- v. second-species thing for point sets from the previous §. There's actually also some technical stuff involving criteria for 'dense' v. 'everywhere-dense' sets that we're omitting, but in essence the way Cantor conceives and generates the transfinite numbers is this:[24]

Assume that P is a second-species infinite point-set. Cantor shows that P's first derived set, P', can be "decomposed"[25] or broken down into the union of two different subsets, Q and R, where Q is the set of all points belonging to first-species derived sets of P', and R is the set of all points that are contained in *every single* derived set of P', meaning R is the set of just those points that all the derived sets of P' have in common. Why not take a second and read that last sentence over again.[26] R is the important part, and it's actually how Cantor first defines 'intersection' for sets, here via the infinite sequence of derived sets P', P'', P''', . . . (the sequence being infinite because P is a second-species set). Unlike our '$\cap$,' Cantor's symbol for intersection is a strange ultracursive '$\mathfrak{D}$'. (Again, we're not going to be doing everything in this much detail.) So the official definition of R is: $R = \mathfrak{D}\,(P', P'', P''', \ldots)$,

[24] **IYI** FN 14's apologies and reassurances pertain to the following text ¶, too.

[25] **IYI** = Cantor's term, which apparently in German doesn't have the postmortem connotations it does for us.

[26] **IYI** Cantor's proof that $P' = Q \cup R$ is emetically complex but wholly valid; please take it on faith.

which, together with the def. of 'second-species set,' means that both the following are true:

$$(1)\quad R = \mathfrak{D}(P^{(2)}, P^{(3)}, P^{(4)}, P^{(5)}, \ldots) \ldots$$
$$(2)\quad R = \mathfrak{D}(P^{(n)}, P^{(n+1)}, P^{(n+2)}, P^{(n+3)}, \ldots)$$[27]

EMBEDDED MINI-INTERPOLATION

What (1) and (2) together really are is a type of proof, the other really famous one besides the reductio. This one's called *mathematical induction.* To prove some statement C_n for all ($n = \infty$) cases by math induction, you (a) prove that C_1 is true for the first case $n = 1$, then (b) assume that C_k is true for the first k cases (you don't know what number k is, but from step (a) you know that it exists—if nothing else, it's 1), and then (c) prove that C_{k+1} is true for the first $(k + 1)$ cases. Weird-looking or no, (a)–(c) ensure that C_n will be true no matter what n is, that is, that C is a genuine theorem.

END E.M-I.

What Cantor's (1) and (2) allow him to do is to define R, as taken from P, as: $R = P^{\infty}$—that is, R is the ∞th derived set of P. And since (again) P is a second-species set, there is no chance that $P^{\infty} = \emptyset$, which means that P^{∞} will itself yield the derived set $P^{(\infty+1)}$, which latter will yield the derived set $P^{(\infty+2)}$, and so on, except here 'and so on' means we can keep generating infinite derived sets of the abstract form[28]

[27] **IYI** Notice how (1)'s got ellipses outside the right paren, too, meaning the sequence continues beyond the finite-superscripted progression. (2)'s ellipses are 100% intraparenthetical because n itself is infinite. Make sense?

[28] The following actual notation is of course **IYI**—but if nothing else it's quite pretty.

$P^{(n_0\infty^\nu + n_1\infty^{\nu-1} + \cdot \cdot \cdot + n_\nu)}$. And since this formula's n and ν are variables, Cantor can construct the following infinite sequence of infinite sets: $P^{(n\infty^\infty)}$, $P^{(\infty^{\infty+1})}$, $P^{(\infty^{\infty+n})}$, $P^{(\infty^{n\infty})}$, $P^{(\infty^{\infty^n})}$, $P^{(\infty^{\infty^\infty})}$, ... Of this sequence he says "We see here a dialectic generation of concepts, which leads further and further, and thus remains in itself necessarily and consequently free of any arbitrariness" . . . by which he means that these "concepts" are real math entities—*transfinite numbers*—rigorously established by the Bolzano-Weierstrass Theorem, G. C.'s own definitions of 'real number' and 'derived set' and 'intersection,' and mathematical induction.

If you object (as some of us did to Dr. Goris) that Cantor's transfinite numbers aren't really numbers at all but rather sets, then be apprised that what, say, '$P^{(\infty^{\infty+n})}$' really is is a symbol for the number of members in a given set, the same way '3' is a symbol for the number of members in the set {1, 2, 3}. And since the transfinites are provably distinct and compose an infinite ordered sequence just like the integers,[29] they really are numbers, symbolizable (for now) by Cantor's well-known system of *alephs* or 'ℵ's.[30] And, as true numbers,

[29] **IYI** In truth there's a much better analogy for transfinite numbers than the integers, viz. some other kind of number that can't actually be named or counted but can nevertheless be abstractly generated—by, say, drawing the diagonal of the Unit Square, or taking the square root of 5, or describing a particular Dedekindian *A* and *B* and interstitial *schnitt.* Regarding all of which please see or await the discussion two text ¶s down.

[30] **IYI** That the *aleph* is a Hebrew letter is sometimes made much of by historians w/r/t Cantor's ethnicity or kabbalistic leanings. More plausible explanations for Cantor's choice of 'ℵ' are that (1) he wanted a whole new symbol for a whole new kind of number, and/or (2) all the good Greek letters were already taken.

transfinites turn out to be susceptible to the same kinds of arithmetical relations and operations as regular numbers—although, just as with 0, the rules for these operations are very different in the case of $\aleph$s and have to be independently established and proved.

(**IYI** We won't be doing a whole lot with them, but if you're curious, here are some of the standard theorems for the addition, multiplication, and exponentiation of transfinite numbers, all either derived or suggested by Cantor. (Please note that sums and/or products of infinitely many terms here have nothing to do with the sums/limits of infinite series in analysis, which series are now known, post-Cantor, as *quasi-infinite.*) Assume that n is any finite integer, and that we've got two distinct transfinite numbers, designated '$\aleph_0$' and '$\aleph_1$,' where $\aleph_1 > \aleph_0$,[31] in which case the following are all true:

(1) $1 + 2 + 3 + 4 + \cdots + n + \cdots = \aleph_0$
(2) $\aleph_0 + n = \aleph_0$
(3) $\aleph_0 \times n = \aleph_0$
(4) $\aleph_0 + \aleph_0 + \aleph_0 + \cdots = \aleph_0 \times \aleph_0 = (\aleph_0)^2 = \aleph_0$
(5) $\aleph_1 + n = \aleph_1 + \aleph_0 = \aleph_1$
(6) $\aleph_1 \times n = \aleph_1 \times \aleph_0 = \aleph_1$
(7) $\aleph_1 + \aleph_1 + \aleph_1 + \cdots = \aleph_1 \times \aleph_0 = \aleph_1$
(8) $\aleph_1 \times \aleph_1 = (\aleph_1)^2 = (\aleph_1)^n = (\aleph_1)^{\aleph_0} = \aleph_1$

N.B. that subtraction and division are possible only in certain miscegenated cases—e.g., for a finite n, $\aleph_0 - n = \aleph_0$; $\frac{\aleph_0}{n} = \aleph_0$—not between transfinites per se. (Again, this is not all that different from the arithmetic of 0.) Note also that the case of

[31] **IYI** re which proofs are on the way in §7c ff.

transfinite exponents like $2^{\aleph_0}$, $2^{\aleph_1}$, etc. is special and gets discussed at length later on. **END IYI**)

In case you're wondering what any of this has to do with the other big ∞-related issues—namely the comparative infinities of the rational numbers v. the real numbers, and surds' role in the continuity of the Real Line—be apprised that one of Cantor's favorite arguments for transfinite numbers' reality[32] is their mathematical/metaphysical similarity to irrationals, which latter Dedekind has already successfully defined in terms of infinite sets. How Cantor puts it is:

> The transfinite numbers themselves are in a certain sense *new irrationals,* and in fact I think the best way to define the *finite* irrational numbers is entirely similar[33]; I might even say in principle it is the same as my method for introducing transfinite numbers. One can absolutely assert: the transfinite numbers *stand or fall* with the finite irrational numbers; they are alike in their most intrinsic nature; for the former like these latter are definite, delineated forms or modifications of the actual infinite.

What's interesting is that this clear, unequivocal statement appears in the same "Contributions to the Study of the Transfinite" where all the way back in §3a we saw Cantor quote and credit St. Thomas's objection to infinite numbers qua infinite sets. Nevertheless, Cantor's own #1 argument for transfinite numbers—an argument repeated in many forms from 1874 to the late 1890s—is that "their existence is confirmed directly by abstraction from the existence of infinite

[32] **IYI** mathematically speaking.

[33] He's talking here about Dedekind's *schnitt* method, which after he got going on ∞ Cantor preferred over his own approach, for rather obvious reasons.

sets."[34] Thus the central project of Cantor's 1874–84 work is to develop a coherent, consistent theory of infinite sets—and please notice the plural 'sets,' because for such a theory to be nontrivial there needs to be more than one kind (meaning mathematical kind, meaning basically size[35]) and some set of rules for evaluating and comparing them.

[34] Because of space considerations we're not going to harp too much on this, but let's emphasize once more here that G. Cantor is, like R. Dedekind, a mathematical Platonist; i.e., he believes that both infinite sets and transfinite numbers really exist, as in metaphysically, and that they are "reflected" in actual real-world infinities, although his theory of these latter involve Leibnizian monads and is best steered clear of. As it happens, Cantor develops all sorts of theological positions and arguments respecting ∞, too, some of which are cogent and powerful and others merely eccentric. Still, as a mathematician and rhetor Cantor is smart enough to argue that one doesn't need to accept any particular metaphysical premises in order to admit infinite sets or their abstract numbers into the domain of legit mathematical concepts. See e.g. this passage from Cantor's prenominate "Contributions . . .":

> In particular, in introducing new numbers, mathematics is only obliged to give definitions of them, by which such a definiteness and, circumstances permitting, such a relation to the older numbers are conferred upon them that in given cases they can definitely be distinguished from one another. As soon as a number satisfies all these conditions, it can and must be regarded as existent and real in mathematics. Here I perceive the reason why one has to regard the rational, irrational, and complex numbers as being just as thoroughly existent as the finite positive integers.

That last and clearest sentence is a tiny blown kiss to L. Kronecker. The rest is obscure enough that J. Dauben gives it the following gloss: "For mathematicians, only one test was necessary: once the elements of any mathematical theory were seen to be consistent, then they were mathematically acceptable. Nothing more was required."

[35] **IYI** Here the best supporting scholium is from Cantor's follower A. A. Fraenkel: "The concept of transfinite magnitude is insignificant so long as only one such magnitude was known to exist."

§7c. This obviously segues into the question whether the Real Line's continuity means that the infinite set of all real numbers is somehow > the infinite set of all rational numbers. To make a very long story short, Cantor's work on this problem proceeds more or less simultaneously with his development of the derived-set and transfinite-number stuff.[36]

All right. In trying to find some way to compare the sizes of two sets that are both infinitely large, Cantor hits on the very concept that's now used in 4th grade to define equality between two sets, namely one-to-one correspondence, or '1-1C'. (Actually 'hits on' isn't quite right, since we've seen both Galileo and Bolzano use one-to-one correspondence to establish their respective paradoxes (though after Cantor's theory they'd be paradoxes no more).) One-to-one correspondence is, as you may already know, the way to establish whether two collections are equal without having to count them. Textbooks use all kinds of different scenarios to illustrate how the 1-1C matchup works, e.g. the fingers on your right v. left hands, the number of patrons v. available seats in a theater, a restaurant's cups v. saucers. Dr. Goris's own chosen trope (which was clearly audience-tailored) involved the numbers of boys v. girls at a dance and having everybody couple up and dance and seeing whether anyone was left standing stricken and alone against the wall. You get the idea. A couple formal definitions: There is a *one-to-one correspondence* between sets *A* and *B* if and only if there exists a method (which technically no one has to know about) of pairing off the members of *A* w/ the members of *B* so that each *A*-member is paired with

[36] **IYI** Really more like 'in conjunction' than 'simultaneously,' since the two projects end up being connected in all sorts of high-level ways.

exactly one *B*-member and vice versa. Sets *A* and *B* are defined as having the same *cardinal number* (a.k.a. *cardinality*) if and only if there is indeed a 1-1C between them.[37]

Now, for the next definition please recall how Galileo generated his eponymous Paradox in §1d. It would also be helpful to remember the formal def. of *subset* one § back. A set *A* is a *proper subset* of set *B* if and only if *A* is a subset of *B* and there is at least one member of *B* that is not a member of *A*.[38] So, by definition, every set is a subset of itself, but no set is a proper subset of itself. Make sense? It ought to, at least for sets with a finite number of members.

But what G. Cantor posits as the defining formal property of an *infinite set* is that such a set can be put in a 1-1C with at least one of its proper subsets. Which is to say that an infinite set can have the same cardinal number as its proper subset, as in Galileo's infinite set of all positive integers and that set's proper subset of all perfect squares, which latter is itself an infinite set.

This feature makes the whole idea of comparing the quote-unquote 'sizes' of infinite sets look very freaky, since by definition an infinite set can have the same size (or cardinality) as a set it's by definition bigger than. What Cantor does here[39] is

[37] **IYI** If you've ever run across references to Cantor's *transfinite cardinals*, this is what they are—the cardinal numbers of infinite sets.

[38] **IYI** Galileo's Paradox rests squarely if covertly on this def. See, for instance, §1d p. 39's "It's also obvious that while every perfect square (viz. 1, 4, 9, 16, 25, . . .) is an integer, not every integer is a perfect square."

[39] **IYI** 'Here' = mostly two seminal articles in 1874 and '78, though he also spends a lot of time fleshing the idea out in his later, more discursive papers. If you'd like the title of the '74 monograph, it's "*Über eine Eigenschaft des Inbegriffes aller reellen algebraischen Zahlen.*" This translates to something like "On a Characteristic of the Set of All Real Algebraic Numbers," regarding which feel free to see FN 53 below.

take yet another element of Galileo's Paradox and turn it into an extremely powerful and important tool for comparing ∞-type sets. This, if you want to keep track, is his first stroke of incredible, nape-tingling genius, although it may not look like much at first. It's the idea of one-to-one correspondence with the set of all positive integers, viz. {1, 2, 3, . . .}. The reason this is critical is that the set of all positive integers can, in principle, be *counted*[40]—as in it's possible to go 'Here's the first member, 1, and the next member is 2, and . . .,' etc., even though as a practical matter the process never ends. Anyway, hence Cantor's concept of *denumerability*: An infinite set *A* is *denumerable* if and only if there is a 1-1C between *A* and the set of all positive integers.[41]

The set of all positive integers also establishes a sort of baseline cardinal number for infinite sets; it's this set's cardinality that Cantor symbolizes w/ his famous '$\aleph_0$'.[42] The idea is that other infinite sets' cardinalities can be evaluated via this baseline cardinal—that is, you can compare them to $\aleph_0$ by seeing whether they can be put in one-to-one correspondence with the positive integers. Here's an example (it isn't Cantor's per se, but it's a good warmup):

Consider whether the set *C* of all positive integers and the set *D* of all integers (incl. 0 and the negatives) have the same

[40] **IYI** This is, in point of fact, what *counting* any collection of *n* things is: it's putting the things in a one-to-one correspondence with the set of integers {1, 2, 3, . . ., *n*}. The equivalence of counting and 1-1C-ing-with-{all integers} is what made set theory the basis for teaching little kids arithmetic in the New Math.

[41] **IYI** N.B. that in Cantorian set theory *denumerable* is related to but not synonymous with *countable*. Def.: A set is *countable* if and only if it is either (a) finite or (b) denumerable.

[42] This traditionally gets called either *aleph-null* or *aleph-nought*.

cardinality. The problem is that there's a crucial difference between these two sets: C has a very first (meaning smallest) member, namely 1, whereas D (which is basically the set $\{\ldots, -n, \ldots, 0, \ldots, n, \ldots\}$) doesn't. And initially it's hard to see how we can test two sets for 1-1C if one of them doesn't have a first member. Luckily what we're talking about here is cardinality, which has nothing to do with the specific order of the sets' members[43]; thus we can futz with the order of set D in such a way that even though D doesn't have a smallest member it does indeed have a first member, here let's say 0. And this single bit of futzing lets us set up, and represent schematically, a perfect 1-1C—

$$\begin{array}{cccccccccccc} C = 1 & 2 & 3 & 4 & 5 & \cdots & n_{\text{even}} & \cdots & n_{\text{odd}} & \cdots \\ \updownarrow & \updownarrow & \updownarrow & \updownarrow & \updownarrow & & \updownarrow & & \updownarrow & \\ D = 0 & -1 & 1 & -2 & 2 & \cdots & (-1)\frac{n}{2} & \cdots & \frac{n-1}{2} & \cdots \end{array}$$

—that proves C and D have the same cardinality. Notice that even though you can never literally finish the matching process with infinite sets, as long as you can establish a procedure for one-to-one correspondence that works for the 1st, nth, and $(n + 1)$th cases, you have proved by mathematical induction that the correspondence will obtain all the way through both sets. In the above example, we've proved that the set of all integers is denumerable even though we can't possibly count every member.[44] The proof's method is © G. Cantor, and the big thing to see is that he is once again able to take an implicit property of something—here math induction's ability to

[43] **IYI** Order starts mattering only with Cantor's *transfinite ordinals*, which enter play in §7g.

[44] **IYI** Notice that we're starting to be able to answer Russell's epigraphic query at the start of §7.

abstract a finite number of results over a ∞ of possible cases—and to make it explicitly, rigorously applicable to infinite sets.

OK, so now it's clear how Cantor can do a size-comparison on the infinite sets of all rational numbers and all real numbers:[45] he can see whether either or both are denumerable. What follow are a series of very famous proofs, most worked out in correspondence with R. Dedekind, published in the 1870s, and then revised and expanded in the early '90s. First the rationals.[46] When you consider the infinite density that Zeno had exploited merely in the geometric rationals between 0 and 1, it appears as if the set of all rational numbers can't possibly be denumerable. Not only does it lack a smallest member, but there isn't even a next-largest member after any given rational (as we've seen two different proofs of). What Cantor notices, though, is that by ignoring 'relations of magnitude' between successive members, we can actually arrange the set of all rationals in a row, something like the row of all positive integers; and in that row there'll be a first member r_1, a second member r_2, and so on. It just so happens that the technical term for putting a set into such a

[45] Command Decision: From here on out, when we talk about 'all numbers' we're going to deal only with positive values. This includes the integers, since after all we've just proved that the set of all integers' cardinality = that of the set of all positive integers. Plus it ought to be obvious that Cantor's proofs for all positive rationals, all positive reals, etc. will still be valid if the relevant infinite sets are doubled to comprise negative values. If you're dubious, then observe that doubling something is the same as multiplying it by 2, and that 2 is finite, and that—by transfinite theorems (3) and (6) in §7b—any ℵ times a finite n will still = ℵ.

[46] **IYI** Again, please be advised that we're doing these proofs in the order that yields the clearest and most logical exposition. N.B. also that there are also some cardinals-v.-ordinals distinctions that for now we're going to fudge.

row is giving a *denumeration* of the set—plus the row itself is called the set's *denumeration*—meaning that here the valid construction of an ordered row will constitute a proof that the set of all rationals really is denumerable (that is, 1-1C-able with, and so equivalent in cardinality to, the set of all integers). Cantor's construction, which is sometimes referred to incorrectly as his 'Diagonal Proof,'[47] runs more or less thus:

As we saw in §6c, all rational numbers can be put in the ratio-of-integers form $\frac{p}{q}$. So we make a 2D array of all these $\frac{p}{q}$, where the top horizontal row is all the rationals of the form $\frac{p}{1}$ (i.e., the integers), and the first vertical column is all the rationals of the form $\frac{1}{q}$, and every rational $\frac{p}{q}$ will be located in the qth row and pth column, like so:

$$
\begin{array}{ccccccccccc}
1 & 2 & 3 & 4 & 5 & 6 & 7 & \cdots & \frac{p}{1} & \cdots \\
\frac{1}{2} & \frac{2}{2} & \frac{3}{2} & \frac{4}{2} & \frac{5}{2} & \frac{6}{2} & \frac{7}{2} & \cdots & \frac{p}{2} & \cdots \\
\frac{1}{3} & \frac{2}{3} & \frac{3}{3} & \frac{4}{3} & \frac{5}{3} & \frac{6}{3} & \frac{7}{3} & \cdots & \frac{p}{3} & \cdots \\
\frac{1}{4} & \frac{2}{4} & \frac{3}{4} & \frac{4}{4} & \frac{5}{4} & \frac{6}{4} & \frac{7}{4} & \cdots & \frac{p}{4} & \cdots \\
\frac{1}{5} & \frac{2}{5} & \frac{3}{5} & \frac{4}{5} & \frac{5}{5} & \frac{6}{5} & \frac{7}{5} & \cdots & \frac{p}{5} & \cdots \\
\frac{1}{6} & \frac{2}{6} & \frac{3}{6} & \frac{4}{6} & \frac{5}{6} & \frac{6}{6} & \frac{7}{6} & \cdots & \frac{p}{6} & \cdots \\
\frac{1}{7} & \frac{2}{7} & \frac{3}{7} & \frac{4}{7} & \frac{5}{7} & \frac{6}{7} & \frac{7}{7} & \cdots & \frac{p}{7} & \cdots \\
\vdots & \vdots & \vdots & \vdots & \vdots & \vdots & \vdots & \vdots & \vdots & \vdots \\
\frac{1}{q} & \frac{2}{q} & \frac{3}{q} & \frac{4}{q} & \frac{5}{q} & \frac{6}{q} & \frac{7}{q} & \cdots & \frac{p}{q} & \cdots
\end{array}
$$

[47] **IYI** What Cantor really calls 'Diagonalization' is his method for proving the nondenumerability of the reals, which as we'll see is quite different.

Granted, a 2D array is not the same as the single ordered sequence/row of true denumeration, but Cantor figures out how to sequence the array's rationals via a single continuous zigzaggy line, like so: start at 1 and go due east one place to 2, then diagonally southwest to $\frac{1}{2}$, then due south to $\frac{1}{3}$, then diagonally northeast to the first row again and 3, then east to 4, then southwest all the way to $\frac{1}{4}$, south to $\frac{1}{5}$, northeast to 5, and so on, as in:

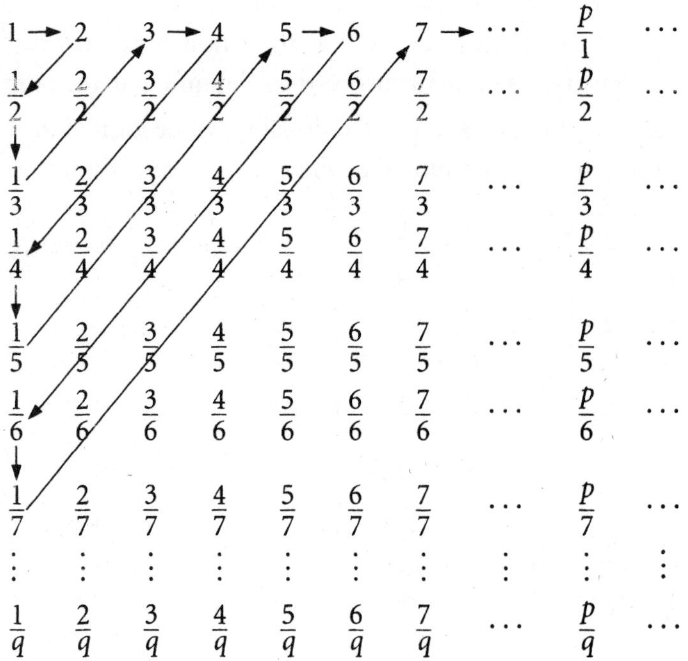

$$\begin{array}{ccccccccc}
1 \rightarrow 2 & & 3 \rightarrow 4 & & 5 \rightarrow 6 & & 7 \rightarrow \cdots & \frac{p}{1} & \cdots \\
\frac{1}{2} & \frac{2}{2} & \frac{3}{2} & \frac{4}{2} & \frac{5}{2} & \frac{6}{2} & \frac{7}{2} \quad \cdots & \frac{p}{2} & \cdots \\
\frac{1}{3} & \frac{2}{3} & \frac{3}{3} & \frac{4}{3} & \frac{5}{3} & \frac{6}{3} & \frac{7}{3} \quad \cdots & \frac{p}{3} & \cdots \\
\frac{1}{4} & \frac{2}{4} & \frac{3}{4} & \frac{4}{4} & \frac{5}{4} & \frac{6}{4} & \frac{7}{4} \quad \cdots & \frac{p}{4} & \cdots \\
\frac{1}{5} & \frac{2}{5} & \frac{3}{5} & \frac{4}{5} & \frac{5}{5} & \frac{6}{5} & \frac{7}{5} \quad \cdots & \frac{p}{5} & \cdots \\
\frac{1}{6} & \frac{2}{6} & \frac{3}{6} & \frac{4}{6} & \frac{5}{6} & \frac{6}{6} & \frac{7}{6} \quad \cdots & \frac{p}{6} & \cdots \\
\frac{1}{7} & \frac{2}{7} & \frac{3}{7} & \frac{4}{7} & \frac{5}{7} & \frac{6}{7} & \frac{7}{7} \quad \cdots & \frac{p}{7} & \cdots \\
\vdots & \vdots & \vdots & \vdots & \vdots & \vdots & \vdots \quad \vdots & \vdots & \vdots \\
\frac{1}{q} & \frac{2}{q} & \frac{3}{q} & \frac{4}{q} & \frac{5}{q} & \frac{6}{q} & \frac{7}{q} \quad \cdots & \frac{p}{q} & \cdots
\end{array}$$

The points on the above line will compose the sequence 1, 2, $\frac{1}{2}, \frac{1}{3}, \frac{2}{2}$, 3, 4, $\frac{3}{2}, \frac{2}{3}, \frac{1}{4}, \frac{1}{5}, \frac{2}{4}, \frac{3}{3}, \frac{4}{2}$, 5, 6, $\frac{5}{2}, \frac{4}{3}, \frac{3}{4}, \frac{2}{5}, \frac{1}{6}, \frac{1}{7}, \frac{2}{6}, \ldots$, from which we can licitly cancel out all the ratios in which p and q

have a common factor, so each different rational appears just once in its most basic form. This process of elimination/reduction then yields the linear sequence 1, 2, $\frac{1}{2}$, $\frac{1}{3}$, 3, 4, $\frac{3}{2}$, $\frac{2}{3}$, $\frac{1}{4}$, $\frac{1}{5}$, 5, 6, $\frac{5}{2}$, $\frac{4}{3}$, $\frac{3}{4}$, $\frac{2}{5}$, $\frac{1}{6}$, $\frac{1}{7}$, . . ., $\frac{p}{q}$, . . ., which sequence constitutes the ordered row required for denumeration,[48] meaning that the set of all rationals is indeed denumerable and therefore has the same cardinality as the set of integers, namely good old $\aleph_0$.

The true Diagonal Proof appears in Cantor's answer to the question whether the set of all real numbers is > the set of all rationals. It should now be obvious that Cantor's proof here will concern the denumerability of the real numbers; i.e., if the reals are denumerable then their cardinality = that of the rationals, and if they're not then the reals are > the rationals. The overall proof is a reductio, and its method of Diagonalization is now regarded as one of the most important proof-techniques in all of set theory. Two preliminary things to note here. (1) Cantor's first, Dedekind-informed, '73–'74 proof of the nondenumerability of the reals involves limits of sequences w/r/t 'nested intervals' on the Real Line and is just hideously complex. The proofs we're doing here are Cantor's revised versions, c. 1890; they are both simpler and more significant than the early one. (2) Notice once again in the following how Cantor uses the decimal form of real numbers and exploits §2c's fact that .999 . . . = 1.0 in order to

[48] **IYI** If this schematic def. seems to you insufficient and you want an actual 1-1C between the sets of rationals and integers, then simply take the prenominate ordered row and match its first member with 1, its second member with 2, . . . and so on and so forth.

represent not just the irrationals but all real numbers as nonterminating decimals—as in for example 0.5 = 0.4999 . . ., 13.1 = 13.0999, etc. This move (which was actually Dedekind's suggestion) ensures that there's only one licit representation of each decimal; we'll see in a minute why Cantor needs to set the real numbers up this way.

So here's the proof. Because it's a reductio, we first assume that the set of all real numbers truly is denumerable—i.e., that it is listable in an ordered row or sequence.[49] This sequence will consist in an infinite table of infinite nonterminating decimals, which table we can show at least the start of, like so:

1st Real # $= X_1.a_1a_2a_3a_4a_5a_6a_7\ldots$
2nd Real # $= X_2.b_1b_2b_3b_4b_5b_6b_7\ldots$
3rd Real # $= X_3.c_1c_2c_3c_4c_5c_6c_7\ldots$
4th Real # $= X_4.d_1d_2d_3d_4d_5d_6d_7\ldots$
5th Real # $= X_5.e_1e_2e_3e_4e_5e_6e_7\ldots$
6th Real # $= X_6.f_1f_2f_3f_4f_5f_6f_7\ldots$
⋮
And so on. . . .

In this table, the *X*'s denote any and all pre-decimal-point integers, and *a*'s, *b*'s, etc. represent the infinite sequences of digits after the decimal points; and the proof's assumption is

[49] **IYI** You may well be able to anticipate some familiar complications here w/r/t what can possibly be the sequence's very first real number, and to see why none of the previous kinds of fiddling with the row-orders will work with the real numbers. In which case, for your own peace of mind, be now advised that Prof. E. Zermelo's famous set-theoretic Axiom of Choice (which is its own briarthicket—see §7f) ensures that we can always construct an ordered set of real numbers in such a way that it's got a bona fide first member. It would be in your interest to just swallow this for now.

that the infinite version of such a table will be exhaustive of *all* the real numbers. This means that the reductio's desired contradiction will require us to prove that such a table does not really exhaust the set of all reals, which proof requires that we come up with a real number that isn't—can't be—included in the table.

What Cantor's Diagonal Proof does is generate just such a number, which let's call *R*. The proof is both ingenious and beautiful—a total confirmation of art's compresence in pure math. First, have another look at the above table. We can let the integral value of *R* be whatever *X* we want; it doesn't matter. But now look at the table's very first row. We're going to make sure *R*'s first post-decimal digit, *a*, is a different number from the table's a_1. It's easy to do this even though we don't know what particular number a_1 is: let's specify that $a = (a_1 - 1)$ unless a_1 happens to be 0, in which case $a = 9$. Now look at the table's second row, because we're going to do the same thing for *R*'s second digit *b*: $b = (b_2 - 1)$, or $b = 9$ if $b_2 = 0$. This is how it works. We use the same procedure for *R*'s third digit *c* and the table's c_3, for *d* and d_4, for *e* and e_5, and so on, ad inf. Even though we can't really construct the whole *R* (just as we can't really finish the whole infinite table), we can still see that this real number $R = X.abcdefghi\ldots$ is going to be demonstrably different from every real number in the table. It will differ from the table's 1st Real in its first post-decimal digit, from the 2nd Real in its second digit, from the 3rd Real in its third digit, . . . and will, given the Diagonal Method here,[50] differ from the table's *N*th Real in

[50] So the reason for the 'Diagonal' thing is the first-digit-in-first-row, second-digit-in-second-row, 45°-angle construction of *R*.

its *n*th digit. Ergo *R* is not—cannot be—included in the above infinite table; ergo the infinite table is not exhaustive of all the real numbers; ergo (by the rules of reductio) the initial assumption is contradicted and the set of all real numbers is not denumerable, i.e. it's not 1-1C-able with the set of integers. And since the set of all rational numbers *is* 1-1C-able with the integers, the set of all reals' cardinality has got to be greater than the set of all rationals' cardinality. Q.E.D.*

***QUICK FOREST-V.-TREE INTERPOLATION**

Let's step back and reflect for just a second here on how stratospherically abstract all this is. And on why set theory, which is arguably the most fundamental part of modern math, is also the most mindbending. Set theory is 100% trivial as long as you're dealing with finite sets, because all relations between such sets can be determined empirically—you just count up their members. In real set theory, we're dealing with abstract aggregates of abstract entities so numerous they cannot ever be counted or completed or even comprehended . . . and yet we are *proving*, deductively and thus definitively, truths about the makeup and relations of such things. In the heat of all this proof and explication, it's easy to lose sight of the utter strangeness of infinite sets, a strangeness which is diminished not one bit by Cantor and Dedekind's having shown that these ∞s lie at the very taproot of math and are required for handling something as basic as a straight line. Apropos this strangeness, here is a nice quotation from philosophers P. Benacerraf and H. Putnam:

> There are the sets; beautiful, imperishable, multitudinous, intricately connected. They toil not, neither do they spin.

> Nor, and this is the rub, do they interact with us in any way. So how are we supposed to have epistemological access to them? To answer, 'by intuition,' is hardly satisfactory. We need some account of how we can have knowledge of these beasties.

—and one from the hardass Intuitionist H. Poincaré:

> A reality completely independent of the spirit that conceives it, sees it, or feels it, is an impossibility. A world so external as that, even if it existed, would be forever inaccessible to us.

—and a rather delicious rebuttal from the Platonist K. Gödel:

> Despite their remoteness from sense experience, we do have something like a perception also of the objects of set theory, as is seen from the fact that the axioms force themselves upon us as being true. I don't see any reason why we should have less confidence in this kind of perception, i.e. in mathematical intuition, than in sense perception, which induces us to build up physical theories and to expect that future sense perceptions will agree with them. . . .

END Q. F.-V.-T. I. RETURN TO §7c AT THE ¶ ON p. 256 W/ ASTERISK AT END

Some addenda regarding these first two proofs. (1) Since the cardinal number of denumerable sets is $\aleph_0$, it looks as if it would make sense to signify the set of all reals' cardinality by '$\aleph_1$'; but for complicated reasons Cantor designates this set's cardinal number ***c***, which he also calls "the power of the Continuum," since it turns out to be the nondenumerability of the reals that accounts for the continuity of the Real Line. What this means is that the ∞ of points involved in continuity is greater than the ∞ of points comprised by any kind of discrete sequence, even an infinitely dense one. (2) Via his Diagonal Proof that $c > \aleph_0$, Cantor has succeeded in characterizing arithmetical continuity entirely in terms of order, sets, denumerability, etc. That is, he has characterized it 100% abstractly, without reference to time, motion, streets, noses, pies, or any other feature of the physical world—which is why Russell credits him with 'definitively solving' the deep problems behind the Dichotomy.[51] (3) The D.P. also explains, with respect to Dr. G.'s demonstration back in §2e, why there will always be more real numbers than red hankies. And it helps us understand why rational numbers ultimately take up 0 space on the Real Line,[52] since it's obviously the irrational numbers that make the set of all reals nondenumerable. (4) An extension of Cantor's proof helps confirm J. Liouville's 1851 proof that there are an infinite number of transcendental irrationals in any interval on the Real Line. (This is pretty interesting. You'll recall from §3a FN 15 that of the two types of irrationals, transcendentals are the ones like π and e that can't be the roots of integer-coefficient polynomials. Cantor's proof that the reals' ∞ outweighs the rationals' ∞ can be modified to show that it's actually the transcendental irrationals that are nondenumerable and

[51] **IYI** Q.v. the beginning of §5e.

[52] **IYI** The proof of this weird factoid, which was way back on §2e's pp. 89–90, can of course now be freed from the requirement that all the hankies and half-hankies and half-half-hankies be infinitesimally small—we merely invoke §5e(1)'s Weierstrassian proof that $\lim\limits_{n \to \infty} (\frac{1}{2^n}) = 0$.

that the set of all algebraic irrationals has the same cardinality as the rationals,[53] which establishes that it's ultimately the transcendental-irrational-reals that account for the R.L.'s continuity.) (5) Given that the D.P. is a reductio proof and that its quantities are in no way constructible, it should come as no surprise that Prof. L. Kronecker and other proto-Constructivists didn't like it at all (re which there's much more a couple §s down). By all accounts, Kronecker's public campaign against Cantor commenced in earnest with the ***c***-related papers.

§7d. Mathematically, you can probably see what Cantor's next big move is. Having proved with ***c*** that there is a power of ∞ greater than $\aleph_0$, he starts looking for infinite sets whose cardinality might be greater than ***c***. His next major proof (which you'll notice still concerns point sets) is an attempt to show that the 2D plane contains a ∞ of points that's greater than the 1D Real Line's ***c*** in the same way that ***c*** is greater than the Number Line's $\aleph_0$. This is the proof of whose final result Cantor famously wrote to Dedekind "*Je le vois, mais je'n le crois pas*" in 1877.[54] It's known in English as his Dimension Proof. The general idea is to show that the real numbers cannot be put into a 1-1C with the set of points in

[53] **IYI** Historically speaking, the earliest nondenumerability result Cantor could actually prove was that the set of all transcendentals was nondenumerable and that the total set of all rationals + all algebraic irrationals had the same cardinality as the rationals. Q.v. here FN 39 supra and the title of Cantor's '74 paper, which should now make more sense.

[54] **IYI** = "I see it, but I don't believe it" (which is a pithy if unintentional bit of anti-empiricism). Just why he says this in French to a fellow

an *n*-dimensional space, here a plane, and hence that the cardinality of the plane's point set is > the cardinality of the set of all reals. The proof's specific cases are the good old Pythagorean Unit Square and the interval [0,1] on the Real Line. (You will remember from §3 that Bolzano's *P. of the I.* had already suggested in 1850 that [0,1] contained as many points as the whole Real Line, which equivalence Cantor now formally proves in his Dimension paper. Since we've already seen a graphical demonstration of the equivalence in §3c, we'll skip this proof except to point out what you can likely anticipate: Cantor shows that whatever type of Diagonalization you use to create a new real number that's >1 can be duplicated to create a new real number in [0,1].)

For the paper's main Dimension Proof, you sort of have to visualize the Unit Square set up like a Cartesian grid, with numerical coordinates corresponding to each and every point on its plane. Cantor's strategy is to use Diagonalization to show that there are numbers corresponding to these 2D coordinates that cannot be found in the set of all reals. As is clear from his letters to Dedekind, Cantor is sure at the outset that such numbers can be generated, since every geometer from Riemann on had operated under the assumption that any space's dimension (as in 1D, 2D, 3D) was uniquely determined by the number of coordinates required to identify a point in that space.

Except that assumption turns out to be wrong, as Cantor discovers in his attempt to construct decimal sequences of 2D

German isn't clear—it seems to have been a way to emphasize emotion. Cantor's scholarly German, too, often switches into French or Greek for no discernible reason. Perhaps this was SOP.

coordinates that will allow planar points to be compared with real-number decimals. The tricky thing is obviously that planar points are specified by pairs of real numbers and linear points by solo reals, so (harking back to Pythagoras and Eudoxus) Cantor has to devise a way to make the two sets of points commensurable. It takes him three years to figure out how to do this. Again, let all relevant numbers be represented by infinite nonterminating decimals. Take any point (x, y) on the Unit Square; these coordinates are writable:

$$x = 0.a_1a_2a_3a_4a_5a_6a_7\ldots$$
$$y = 0.b_1b_2b_3b_4b_5b_6b_7\ldots$$

which combine to make up the point (x, y)'s unique[55] decimal representation:

$$0.a_1b_1a_2b_2a_3b_3a_4b_4a_5b_5a_6b_6a_7b_7\ldots$$

And to this point there will clearly correspond a unique point z in the R.L. interval [0,1], namely the z that equals the real number $0.a_1b_1a_2b_2a_3b_3a_4b_4a_5b_5a_6b_6a_7b_7\ldots$.[56]

[55] This uniqueness is critical. You can't allow two different decimal ways to represent the same point, because the whole idea is to see whether to each particular point in the U. Square there corresponds a particular point in the R.L.'s [0,1]. It should now be 100% clear why Cantor needs to stipulate that numbers like $\frac{1}{2}$ are to be designated *only* by .4999 . . . and so on. If you remember the mention of this stipulation in §7c, be now apprised that it was this Dimension Proof w/r/t which Dedekind suggested it to Cantor, pointing out the "unique mapping" (which was D. & C.'s original term for 1-1C, in letters) would be screwed up if both .4999 . . . and .5000 . . . were allowed.

[56] **INTERPOLATION THAT IS SO IYI IT'S NOT EVEN IN THE MAIN TEXT** Technically it's a little more complicated than that. But not a lot. Cantor's original Dimension Proof is sort of unnecessarily recondite. It involves defining

So, by straightforward extrapolation from the Unit Square and [0,1], every point on a 2D plane can be put into a 1-1C with a point on the R.L. in just this way, and vice versa. More, Cantor's (relatively) simple method of combining coordinates

---

the decimal representation of point (x, y) as the convergent series $\beta_1 \frac{1}{10} + \beta_2 \frac{1}{10^2} + \beta_3 \frac{1}{10^3} + \cdots + \beta_n \frac{1}{10^n} + \cdots$, and then "pulling the terms of [this series] apart" in such a way as to form for each member a sequence of "ρ independent variables" in [0,1], these latter designated $\alpha_1, \alpha_2, \alpha_3, \alpha_4, \ldots, \alpha_\rho$. The "pulling apart" and subsequent mapping (as well as the reverse, so that the $\alpha \leftrightarrow \beta$ correspondence works both ways) is accomplished via four equations, of which the first looks like: '$\alpha_{1,n} = \beta_{(n-1)\rho+1}$'. Which may not make the 1-1C here terribly obvious. The real proof's rub (as explained in 1979 by the redoubtable Dr. G.) is that the '$a_1 a_2 a_3 / b_1 b_2 b_3$' description of the planar coordinates (x, y) we gave above is a bit too simple in that it makes it look like a and b are individual digits. What Cantor's technique really does is break x and y up into little *chunks* of post-decimal digits; the rule is that each chunk terminates at the very first nonzero digit you hit (which is another, more technical reason why Cantor can't have integers and rationals ending in .0000 . . ..) So say for example that $x = 0.020093089\ldots$ and that $y = 0.702064101\ldots$, in which case they get broken down into:

$x = 0.02\ 009\ 3\ 08\ 9\ldots$
$y = 0.7\ 02\ 06\ 4\ 1\ 01\ldots$

It's these chunks that get combined tit for tat to compose point (x, y)'s unique decimal representation: $(x, y) = 0.02\ 7\ 009\ 02\ 3\ 06\ 08\ 4\ldots$. And here the extra spaces are just illustrative aids; the actual decimal rep. of (x, y) is 0.02700902306084 . . .. This is, of course, also a real number, namely the [0,1] point z that equals 0.02700902306084 . . .. And what's ingenious about the chunks-of-digits device is that if there's a z' that differs from z by even one digit lying n places out past the decimal (which of course there will be—a ∞ of such z's, in fact), then the relevant (x, y) will also be different in the chunk comprising that nth digit, so the relevant correspondence is *biunique*, meaning for each z there's a unique (x, y) and vice versa, which means it's a genuine 1-1C.

into a single real number means that the same general technique can be used to show that a 3D cube, a 4D hypercube, or actually any *n*-dimensional figure's total point-set has the same cardinality as the R.L.'s set of real numbers, namely ***c***. This is an extraordinary result, and it's why Cantor wasn't disappointed at having failed to prove his original premise: he'd discovered an incredible depth and richness to the Continuum, and his proof showed (this is him writing to Dedekind) "what wonderful power there is in the real numbers, since one is in a position to determine uniquely, with a single coordinate, the elements of an *n*-dimensional continuous space."

Cantor's discovery that lines, planes, cubes, and polytopes[57] were all equivalent as sets of points goes a long way toward explaining why set theory was such a revolutionary development for math—revolutionary in theory and practice both. Some of this goes all the way back to the Greeks' commensurability problem and classical calc's ambivalent relationship to geometry. Uneasiness about using quantities like x^2 and x^3 in the same equation (since squares entailed 2D areas and cubes → 3D volumes) had persisted for centuries, and the 1800s' emphasis on rigor made the geometric ambiguities even less palatable. To make a long story short, Cantorian set theory helps unify and clarify math in the sense that all mathematical entities can now be understood as fundamentally the same kind of thing—a set. Plus, in the new non-Euclidean geometries,[58] Cantor's finding that all geometric-point-sets are

[57] **IYI (inserted at editor's insistence)** = the prurient term for polyhedra in 4+ dimensions.

[58] **IYI** as mentioned in §§ 5b and 5d. (And in the latter §, Riemannian geometry's use of ∞ and ∞-related points was at least touched on.)

transfinitely equivalent (i.e., that they all had the cardinality ***c***) is of major importance, particularly in the idea of dimension, as Cantor also observes to Dedekind:

> This [=G.C.'s] view seems opposed to that generally prevailing in particular among the advocates of the new geometry, since they speak of simply infinite, two, three, . . ., *n*-dimensional infinite domains. Sometimes one will even find the idea that the infinity of points of a [2D] surface may be produced so to speak by squaring, that of a solid by cubing the infinity of points of a line.

It goes without saying, though, that our 'revolutionary development' and 'major importance' stuff is in hindsight. As has been abundantly foreshadowed, it's not the case that mainstream math immediately dropped to one knee with arms out to welcome Cantor's post–Uniqueness Theorem proofs. Particularly re the Dimension Proof, mathematicians of nearly all stripe and school lined up to revile it. Besides the general objections in §7c, Constructivists especially hated the idea of somehow creating 1D irrationals out of 2D combinations of other irrationals, as well as the 'noncontinuous mapping' the Dimension Proof produced between line-points and plane-points[59]; and it was actually Cantor's

[59] This gets into a pretty specialized area of function theory, but in essence *noncontinuous mapping* means that if you travel continuously along the points in [0,1] on the R.L., the corresponding points on the Unit Square won't form a continuous curve but will be spread out patternlessly all over the place. (**IYI** With respect to the second ¶ of §7d above, it turns out that Riemann et al.'s assumption was wrong in an interesting way: the dimension of a given set of points depends not only on how many coordinates per point there are, nor even on the cardinality of the total

Dimension-Proof paper[60] that L. Kronecker first intrigued to get rejected from a journal he was on the editorial board of, over which Cantor spent a great many letters venting spleen. But it wasn't just Constructivists or fundamentalists. See, for just one example, these lines from P. Du Bois-Reymond—who is not a Kroneckerian but a mainstream analyst in the Aristotle–Gauss tradition of potential-only ∞s—in a review of the Dimension Proof:

> It seems utterly repugnant to common sense. The simple fact is that this is the outcome of a type of reasoning that allows Idealistic [=Platonic] fictions to assume the role of genuine quantities even though they are not even truly limits of representations of quantities.[61]

§7e. Anyway, so we've established at least and maybe at most two distinct orders of infinite sets, $\aleph_0$ and $\mathbf{c}$,[62] and it's

---

point-set, but also on the particular way the points are distributed. This latter is an issue in *point-set topology*, regarding which all we're in a position to say in this booklet is that it is yet another branch of math that wouldn't exist without Cantor's work on ∞.)

[60] **IYI** date = 1878; title = "*Ein Beitrag zur Mannigfaltigkeitslehre*," or roughly "A Contribution to the Theory of Manifolds/Aggregates/Sets."

[61] **IYI** It's probably obvious that D. B.-R.'s specific "fictions" here are the composite decimal reps. for (x, y); but in the context of the whole review he's also talking about the infinite sets of R.L.- and U.S.-points the decimals are mapping. (N.B. that exactly the same charge could have been leveled against Dedekind's *schnitt* theory's *A* and *B*—for some reason Dedekind never drew the same kind of fire Cantor did.)

[62] Cantor often refers to these as the *first number class* and *second number class*, respectively.

now appropriate to ask what exactly these cardinal numbers have to do with the transfinite numbers that we saw Cantor manufacture out of R and $\mathfrak{D}$ and the derived-sets-of-derived-sets thing in §7b. With the big specific question being whether the infinite sequence of infinite sets $P^{(n\infty^{\infty})}$, $P^{(\infty^{\infty+1})}$, $P^{(\infty^{\infty+n})}$, etc. can be shown to correspond to an infinite hierarchy of greater and greater cardinal numbers, or whether $\aleph_0$ and c are the only infinite cardinals and there are no real ∞s beyond the transdimensional power of the Continuum.

Cantor's next big discovery is that you can validly construct an infinite sequence of infinite sets with larger and larger cardinal numbers using nothing but the formal properties of sets.[63] These properties involve the concepts of the subset and of the *Power Set*, the second of which is hereby defined, for some set A, as simply the set of all subsets of A. Meaning every member of $P(A)$ is some subset of A. This turns out to be heavier than it looks. Every set, finite or not, has a Power Set[64]; but what Cantor's able to prove is that even if set A is infinite, its Power Set $P(A)$ will always have a larger cardinal number than A—more specifically, he's able to prove that the cardinal number of $P(A)$ will always equal 2^A.[65]

[63] **IYI** Textbooks often state this as an abstract theorem, like 'Given any infinite set S, it is possible to construct another infinite set S' with a greater cardinal number'.

[64] **IYI** This principle is known in set theory as the Power Set Axiom. One reason it's an Axiom is that it drops right out of the definitions of 'subset' and 'empty set,' as will be evident just below in the main text. There are problems with the P.S.A., though—q.v. the interpolative §7f below.

[65] **IYI** Technically, this ought to be written '$P(A) = 2^{\bar{A}}$,' where '$\bar{A}$' stands for the cardinal number of A. Having stated this for the record, we'll just write it informally from now on.

And this $A \rightarrow 2^A$ thing ends up being crucial for navigating the transfinite, in which realm it turns out that one sort of jumps, quantumlike, from one number class to the next, with nothing in between: $2^{\aleph_0} = \aleph_1$, $2^{\aleph_1} = \aleph_2$, and so on (as it were).

Cantor's Power Set proofs are extremely intricate, and we have to kind of build up to them. And 4th grade was doubtless a long time ago for everybody, so in case we haven't already done so let's explicitize that the formal way to designate a set is to put its members inside {braces} like this, and that the symbolism for 'item *a* is a member of set *A*' is '$a \in A$'. Let's further remind you that 'subset' is by definition more inclusive than 'proper subset,' and that included among the subsets of any set *A* will be (1) *A* itself and (2) the empty set, symbolized 'Ø' or sometimes '{ }'.[66] Since any set, therefore, has at least some subsets, it follows that every set has a Power Set. To see informally that the number of members of *A*'s Power Set always equals $2^{(\text{Number of Members of } A)}$, let *A* be the three-member set {1, 2, 3}. *A*'s subsets here are: { }, {1}, {2}, {3}, {1, 2}, {1, 3}, {2, 3}, {1, 2, 3}, of which there are exactly 8, or 2^3. A more rigorous way to prove $P(A) = 2^A$ is by mathematical induction, which technically isn't the way Cantor does it but is at least implicit in Cantor's proof; plus it's comparatively easy. Please review or resummon §7b's thing on the three steps of proof by math induction, which here will go like:

(a) Prove that the cardinality of $P(A)$ equals 2^A for a set *A* with just one member. Such an *A* has as subsets the following: Ø and *A* itself, which means its $P(A)$ has two

[66] **IYI** Again, technically it would be better to say that 'Ø' is the symbol for { }, which latter *is* the empty set . . . but you get the idea.

members, which is 2^1 members, which is 2^A members, so bingo.[67] (b) Assume it's true that if *A* has *k* members, P(*A*)'s cardinal number = 2^k. (c) Prove that if *A* has $(k + 1)$ members, $P(A) = 2^{(k+1)}$. From step (b) we know that *A*'s first *k* members yield 2^k subsets of *A*. We now take each one of these 2^k subsets and form a brand new subset that also contains the very last of *A*'s $(k + 1)$ members (i.e., the new, extra member designated by the '+1'). We can form exactly 2^k of these new '+1' subsets—one for each of the original subsets. So now we've got the original 2^k subsets that don't contain the new '+1' member, and we've got 2^k new subsets that *do* contain it. That's $(2^k + 2^k)$ subsets, which is equivalent to (2×2^k) subsets, which equals $2^{(k+1)}$ subsets. So (c) is also proven. So sure enough, $P(A) = 2^A$.

For our purposes, Cantor's got two main Power Set proofs. In neither one is he worried about the 2^A thing yet: what he's basically concerned to show is that even for an infinite set *A*, $P(A) > A$.[68] The first version, which dates around 1891, is important mainly because it shows what a potent reductio-weapon the Diagonalization technique is. It can be considered a proof that the set of all subsets of the set of integers is not denumerable[69]—which, since Cantor's already

[67] **IYI** You can prove $P(A) = 2^A$ for the empty set too. If $A = \emptyset$, it has 0 members. It does, however, have a subset—viz. $\emptyset$, since the empty set is a subset of every set. So here $P(A) = 1$, which is 2^0.

[68] This will make sense if you remember that the overall context of these proofs is Cantor's attempt to derive infinite sets (a.k.a. number classes) whose cardinality exceeds *c*.

[69] **IYI** In point of fact, 1891's is really a proof that this Power Set is uncountable. But recall that a set's being countable requires its being either finite or denumerable, and it's easy to show that the relevant set here isn't

shown that the set of integers *is* denumerable, will obviously mean that its Power Set has a higher cardinal number than $\aleph_0$.

Here's the proof. Call the set of all integers *I*; call *I*'s Power Set P(*I*). We know from §7c that in order for P(*I*) to be denumerable, it has to be possible to set up a one-to-one correspondence between P(*I*) and *I*. The present proof is a reductio, so assume that verily such a 1-1C between P(*I*) and *I* is possible. This (as we also know from §7c's Diagonal Proofs) means that the 1-1C can be charted in an array like the following partial example, with the members of *I* on the left and the subsets of *I* (which are also the members of P(*I*), and can be in any sort of random order we want) on the right:

ARRAY #1

I	P (*I*)
0 ↔	{All Integers}
1 ↔	{ }
2 ↔	{All Even Integers}
3 ↔	{All Odd Integers}
4 ↔	{All Primes}
5 ↔	{All Integers > 3}
6 ↔	{All Perfect Squares}
7 ↔	{All Perfect Cubes}
. ↔	.
. ↔	.
. ↔	.

finite. Since the set of all integers {1, 2, 3, 4, 5, . . .} is itself infinite, we have only to take each individual member, put braces around it, and realize that {1}, {2}, {3}, etc. are each subsets of the set of all integers. So there's no way the set of all such subsets can be finite. So the real issue is whether the Power's Set's denumerable.

As it happens, we can tweak this array's informative range by exploiting a property of its 1-1C that you might already have noticed if you've spent any time thinking about why exactly the relation between sets and their Power Sets is always 2^A rather than 3^A or some other x^A. The deep answer is that the '2' in 2^A reflects a particular kind of decision procedure. For each subset *s* of some set *A*, you have exactly two choices with respect to each member *a* of *A*: either *a* is a member of *s*, or it isn't. That last sentence probably requires more than one read. It's hard to put it clearly in natural language, but the idea itself isn't that complicated. *A* is a set, *a* is some particular member of *A*, *s* is some particular subset of *A*. Ask whether *a* happens to be a member of *s*. Well, either it is or it isn't. You exhaust all the possibilities regarding *a*'s membership in *s* by including *a* in *s* once and excluding *a* from *s* once—thereby producing the duo of subsets *s* and *s′* w/r/t each *a*.

(**SEMI-IYI** Here's one of those places where it's simply impossible to tell whether or not what's just been said will make sense to a general reader. If the abstract is-*a*-a-member-of-*s*-or-not thing is clear enough so that you understand why it alone explains why a set *A* with three members will have 2^3 subsets, feel free to skip the rest of this ¶. If it isn't, we'll do a concrete example. Let's say *A* is the same set {1, 2, 3} unpacked on p. 267, where we listed *A*'s subsets: { }, {1}, {2}, {3}, {1, 2}, {1, 3}, {2, 3}, {1, 2, 3}. Take a look at those subsets and see how many times any particular member of *A*—let's say the member 1—is included in the eight total subsets. You'll notice it's included in four of the subsets and excluded from four. If you look at *A*'s member 2, you'll see it's the same thing: 2 is present in four and absent from four. Same with 3. Can you see why? There are eight total subsets; half of them contain any particular member of *A*, and half of them don't. You can actually construct the set of all *A*'s subsets this way. Take any member of *A*. If your first subset *s* doesn't contain the member, your next one *s′* will. Or obversely. That is, for any particular member and subset, there are two choices, and the set of all subsets will comprise them both. Two choices for each member. Hence the number of subsets of {1, 2, 3} will be $2 \times 2 \times 2$, or 2^3. If this still fails to make the basic idea clear, you're asked to please just eat it (the idea) because this is the best we can do.)

OK, so this means that we can take Array #1 and sort of expand it sideways by asking, for every integer in set *I*, whether it really is part of its corresponding subset in the P(*I*) column, and entering 'Yes' if the integer is in that particular subset and 'No' if it isn't. As in:

ARRAY #2

I	P(*I*)	0	1	2	3	4	5	6	7	...
0 ↔	{All Integers}	Yes	Yes	Yes	Yes	Yes	Yes	Yes	Yes	...
1 ↔	{ }	No	No	No	No	No	No	No	No	...
2 ↔	{All Even Integers}	No	No	Yes	No	Yes	No	Yes	No	...
3 ↔	{All Odd Integers}	No	Yes	No	Yes	No	Yes	No	Yes	...
4 ↔	{All Primes}	No	No	Yes	Yes	No	Yes	No	Yes	...
5 ↔	{All Integers > 3}	No	No	No	No	Yes	Yes	Yes	Yes	...
6 ↔	{All Perfect Squares}	Yes	Yes	No	No	Yes	No	No	No	...
7 ↔	{All Perfect Cubes}	Yes	Yes	No	No	No	No	No	No	...
⋮	⋮	⋮	⋮	⋮	⋮	⋮	⋮	⋮	⋮	⋮

And once this Array #2 is set up, we can easily show that the assumed correspondence between *I* and P(*I*) isn't exhaustive and so isn't a valid 1-1C. We do this by using good old Diagonalization to construct a subset of *I* that will never ever show up in the table's *I* ↔ P(*I*) correspondence. It's the subset defined by starting at the extreme northwest corner of Array #2's 'Yes'/'No' table and going diagonally southeast, changing 'Yes's to 'No's and vice versa throughout—like so:

ARRAY #3

0	1	2	3	4	5	6	7	...
No	Yes	Yes	Yes	Yes	Yes	Yes	Yes	...
No	*Yes*	No	No	No	No	No	No	...
No	No	*No*	No	Yes	No	Yes	No	...
No	Yes	No	*No*	No	Yes	No	Yes	...
No	No	Yes	Yes	*Yes*	Yes	No	Yes	...
No	No	No	No	Yes	*No*	Yes	Yes	...
Yes	Yes	No	No	Yes	No	*Yes*	No	...
Yes	Yes	No	No	No	No	No	*Yes*	...
⋮	⋮	⋮	⋮	⋮	⋮	⋮	⋮	⋱ ⋮

All we know about this new subset is that it includes 1, 4, 6, and 7, and that it differs in at least one member from each subset (a.k.a. each member of P(*I*)) in the original 1-1C. Naturally, our Array #3 is just a fragment, but by continuing the simple process of Diagonal 'Yes'–'No' switching we can guarantee that the new subset generated thereby will differ from all the 1-1C's subsets no matter how far out in the pairings we get. Hence a true 1-1C between *I* and P(*I*) is impossible.

Which means $P(I)$ is nondenumerable,[70] which means its cardinality is greater than $\aleph_0$. Q.E.D.

While there are good reasons why we've gone through the proof in such graphic detail here, be advised that this is not the way G. Cantor does it. The truth is that he never explicitly lays out the Diagonal Proof for $P(I) > I$ and therefore for $P(A) > A$; he merely alludes to it as a "natural extension" of his Diagonal Proof of the nondenumerability of the real numbers.[71] The argument for $P(A) > A$ that he does give is a

[70] **IYI** FROM SERIES EDITOR'S LETTER OF QUERIES ON MS. VERSION OF BOOKLET: "p. 272, paragraph following graphic of 'Array #3': so, in other words, no matter how many subsets of *I* we come up with, we can always create new ones? If so, do you want to say something like that, just to spell it out?" FROM TESTY AUTH. REPLY: "No we do *not* want to say something like that, because it's wrong. What Array #3 shows is that no matter how infinitely or ∞^{∞}ly many subsets of *I* we list, it's provable that there will always be *some* subsets that aren't on the list. This is, recall, what 'nondenumerable' means: incapable of being exhaustively listed/rowed/Arrayed (and, again, it's why there will always be more irrationals than Hankies of Death back in Dr. G.'s §2e demo—hankies, like integers and rationals, can compose only a *denumerable* ∞). Plus the 'we can always create new ones' part is deeply, seriously wrong: we're not *creating* new subsets; we're proving that there *do* exist and *will always* exist some subsets that no list or integral 1-1C can capture. W/ 'exist,' admittedly, requiring a wiseass 'as it were' or 'whatever that means' or something—but the reader'd have to be a radical-Shiite Kroneckerian to believe that what we're doing in this proof is really *creating* these new subsets."

[71] **IYI** This brings up an important issue. You may well have noticed how closely the Diagonal Proof of $P(I)$'s nondenumerability resembles §7c's Diagonal Proof of {all real numbers}'s nondenumerability. And now, given that both $P(I)$'s and {all reals}'s cardinalities are $> \aleph_0$, you may well be wondering whether $P(I)$'s cardinal number is $\boldsymbol{c}$ just as {all reals}'s is. In which case you have derived, on your own, a version of one of the most

bit of a skullclutcher, but it ends up playing a key role in our Story's dénouement and so needs to be spelled out. This proof's wholly abstract and nonspecific, designed to show only that from any infinite set *A* you can construct some infinite set *B* whose cardinal number trumps that of *A*. It would maybe be good to prepare yourself, emotionally, for having to read the following more than once:

A is an infinite set; *B* is the set of all subsets of *A*.[72] Because all sets are by definition subsets of themselves, *A* is a subset of *A*, meaning *A* is a member of *B*; so it's definitely possible to set up a 1-1C between all the members of *A* and at least one member of *B*. It is not possible, however, to set up a 1-1C between all the members of *A* and *all* the members of *B*. We're going to prove this by reductio, so we assume the customary position and posit that such a 1-1C is indeed constructed and is exhaustive of both infinite sets. Now, let *a* be any member of *A* and *b* be any member of *B* (so *b* is any subset of *A*). As we saw with Array #2 above, the 1-1C between *A* and *B* can be wholly random in the sense that, in any individual correspondence $a \leftrightarrow b$, *a* may or may not be a member of the *b* it's paired with. For instance, the integer 3 got paired with {All Odd Integers} and is itself an odd integer, whereas 6 got paired with {All Perfect Squares} but is not a perfect square. It will be the same with our present 1-1C and its infinite pairings $a \leftrightarrow b$: sometimes *a* will be a member of the

---

profound problems in Cantorian set theory, which problem gets hashed out at length in §7g. The point being that you are 100% right to be wondering about P(*I*)'s relation to *c*, but just hang on.

[72] **IYI** meaning that *B* is also P(*A*)—but it's easier if you forget about the whole Power Set thing for this proof.

subset *b* it's paired with; sometimes it won't. This is all fairly straightforward. Now, though, consider the total of all *a*'s in the 1-1C that are *not* members of the *b*'s they're matched with. Let ϕ be the set of all such *a*'s. ϕ is, of course, a subset of *A*, which means that ϕ is a member of *B*—and yet it's provable that ϕ cannot be included in the supposedly exhaustive 1-1C between *A* and *B*. For if ϕ is included, it's matched with some *a*, and there are as we've seen only two options: either this *a* is itself a member of ϕ, or it isn't. If *a* is a member of ϕ—but it can't be, since this contradicts the definition of ϕ. But if *a* is *not* a member of ϕ, then it is, by definition, a member of ϕ—which it can't be, but so it must be, but so it can't be . . . and so you've got your LEM-grade contradiction either way. Hence no true 1-1C between *A* and *B* is possible; hence *B*'s cardinality is > *A*'s cardinality. *Quod erat dem.**

***SEMI-INTERPOLATIVE §7f.**

Please notice the way this last proof resembles the ancient Greek 'I Am Lying' paradox[73] where if the sentence is true it's false and if it's false it's true. Meaning we've now entered the chasmal terrain of self-reference. This is the real reason we just slogged through Cantor's $B > A$ proof—it opens up a whole new kind of crevasse for modern math.

Though it's not strictly in our purview, be informed that in the 1930s Prof. K. Gödel[74] will use something very much like Cantor's '$(a \notin \phi) \rightarrow (a \in \phi)$' device to prove his devastating

[73] **IYI** sometimes a.k.a. *Eublides' Paradox* to distinguish it from Epimenides' 'All Cretans Are Liars' variant—long story.

[74] **IYI** 1906–1978, modern math's absolute Prince of Darkness, referenced all the way back in §1a and elsewhere.

Incompleteness Theorems. (In crude terms, Gödel will prove that certain well-formed math propositions are true and yet unprovable by deriving 'Proposition P: Proposition P is unprovable' as a theorem.) More important for our purposes is this idea that sets can include other sets as members, which is essential to the concept of Power Sets and certainly looks innocent enough . . . except after Cantor's proof it turns out to be to be a veritable swan-dive into the crevasse of self-reference. Example: Consider the theorem Cantor's just proved, that the set of all subsets of set *A* will always contain more members than *A* itself. But suppose now that *A* is defined as 'the set of all sets'. By definition, this *A* will contain all its subsets, since these subsets are sets—so here there's no way $P(A) > A$. Upshot: The same sets-of-sets principle that Cantor needs in order to build a hierarchy of infinite sets yields a paradox almost right away.

Historical evidence shows that Cantor knew about the set-of-all-sets paradox by c. 1895,[75] though never once in his published work does he mention it. It's nevertheless known now as Cantor's Paradox. It's also regarded as the basis for the most famous set-theoretic paradox of all, which is usually called Russell's Antinomy because the ubiquitous B. Russell used it to torpedo G. Frege's *Foundations of Arithmetic* in 1901.[76] We can sketch Russell's Antinomy very quickly and

[75] **IYI** We know, for example, that he told D. Hilbert about it, and it's mentioned in at least one of Cantor's letters to Dedekind. Note now for later that there's also another paradox he stumbled on and also didn't publish, which is known today as the Burali-Forti Paradox and has to do with transfinite ordinals, which as mentioned are themselves upcoming.

[76] **IYI** The Frege-Russell thing is a long but much-loved story among math historians, very easy to find elsewhere. (N.B. that Russell's Antinomy is just

easily because almost everybody's heard of it sometime or other. Though it drops right out of Cantor's abstract Power Set proof, the Antinomy also plays havoc with the main criterion in Cantor's definition of 'set,' which (you'll recall from §7a FN 16) is that there's always a procedure such that for any given item you can always determine whether it's a member of a given set.[77] So here is Russell's Antinomy. As we've seen, some sets are members of themselves and some aren't. Actually, most aren't—as in for example the set of all chairs is not itself a chair, the set of all entities that can tie a knot in a cherry-stem with their tongue cannot itself tie such a knot, etc. But some sets do contain themselves as members, e.g. the set of all sets, the set of all abstractions, the set of all entities that cannot tie a cherry-stem knot. Russell calls a set that is not a member of itself a *normal set*, and one that is self-containing an *abnormal set.* So now consider the set *N* of all

---

as often called Russell's Paradox—but it gets tiresome saying 'paradox' over and over.)

[77] **IYI** Once again, it's all a bit more complicated than that. What Russell's Antinomy really exploits is an unsound axiom in early set theory called (no kidding) the *Unlimited Abstraction Principle,* which in effect states that every conceivable feature/condition determines a set—i.e., that given any conceivable property, there exists a set of all entities possessing this property. Three quick remarks about the U.A.P. (1) Notice its intriguing resemblance to Plato's One Over Many argument from §2a. (2) It ought to become evident soon in the main text why the U.A.P. is faulty and enables Russell's Antinomy. (3) Please hold neocortically for just a few pages the fact that the Zermelo–Fraenkel–Skolem system of axioms for set theory amends the U.A.P. to the *Limited Abstraction Principle,* which holds that given any property *p* and a set *S,* we can form the set of all elements of *S* that have *p.*

normal sets—is *N* a normal set?[78] Well, if *N* is a normal set, then by definition it isn't a member of itself; but *N* is the set of all sets that aren't members of themselves, so *N* *is* a member of itself if it isn't a member of itself; although now if *N* really *is* a member of itself then it can't be a member of the set of all sets that aren't members of themselves, so *N* actually *isn't* a member of itself, in which case it *is* . . . and around and around ad inf.

This kind of paradox, like the '$(a \notin \phi) \rightarrow (a \in \phi)$' cruncher in Cantor's reductio, is officially known as a *Vicious Circle.* The 'vicious' here means roughly the same thing it did in §2a's VIR, namely that it becomes logically impossible to do something we're logically required to do. In VC paradoxes like Russell's and Cantor's, what we cannot do is determine whether something is or isn't a member of a set, which violates both the formal definition of 'set' and (way worse) LEM. So these are not lightweight problems.

By this point you've almost certainly discerned the Story of ∞'s overall dynamic, whereby certain paradoxes give rise to conceptual advances that can handle those original paradoxes but in turn give rise to new paradoxes, which then generate further conceptual advances, and so on. If you're one of the readers who is bothering with the '**IYI**' footnotes, you've already seen stuff about one kind of technical remedy for Russell's Antinomy, which is Zermelo et al.'s replacement of

[78] **IYI** The remainder of this text-¶ is skippable if you can already see how the paradox works just from "Is *N* a normal set?" (**IYI**$_2$ Russell also has a famous way to set up his Antinomy in natural language, to wit: Imagine a barber who shaves all and only those who do not shave themselves—does this barber shave himself or not?)

the Unlimited Abstraction Principle with the Limited A.P. Another type of solution is the prohibition of *impredicative definitions* championed by J. H. Poincaré (1854–1912), a major figure in topology who incidentally was, after Kronecker's death in 1891, the #1 opponent of transfinite math.[79] Poincaré's definition of *impredicative* is somewhat shifty, but in essence it means defining an object in terms of a whole group of objects of which it's a part. Even more essentially, an impredicative definition depends on self-referential properties and descriptions, and 'The set of all sets that are not members of themselves' is a perfect example of such a definition (as are 'The set of all sets' and Cantor's def. of set ϕ in the $B > A$ proof above). This all gets very involved, but Poincaré's general tactic is to characterize impredicative definitions in terms of the paradoxical results they can yield,[80] which then forms the logical argument for disallowing them. It's rather the same way dividing by 0 got outlawed. Unfortunately, the formal definitions of all sorts of terms and concepts in analysis, from 'sequence' and 'series' to 'limit point' and 'lower bound,' are also impredicative—not to mention that the concept of impredicativity can itself be made to generate nasty VCs[81]—so Poincaré's solution never really caught on.

[79] **IYI** In this opposition Poincaré's often associated with the finite-point-set specialists E. Borel and L. Lebesgue, and in the metaphysics of math this trio's sometimes known as the Anti-Platonic School.

[80] very much like the Greek characterization of ∞ as *to apeiron*.

[81] **IYI** Here's one you may already have anticipated: If some quality is impredicative if it applies to itself—say e.g. the quality of being expressible in natural language, or correctly spelled, or abstract—then we can call a quality 'predicative' if it doesn't apply to itself. So this quality of being predicative—is it predicative, or impredicative?

Russell's own proposed way to avoid his and Cantor's eponymous Paradoxes is the Theory of Types, which to make a very long story short is part of Russell's foundational program for trying to show that all math is reducible to symbolic logic. The Theory of Types is a sort of grammar of abstraction that disallows certain kinds of propositions in which different Types of entities are treated as equivalent. Meaning, in essence, metaphysically equivalent.[82] The idea is that sets of individuals are not the same Type of entity as individuals themselves, and sets of sets are not the same Type as sets of individuals, and so on. And a particular entity's Type is a direct function of how abstract that entity is, so you end up with a set-theoretic hierarchy that resembles the informal abstraction-Levels we talked about in §1b—Russell's Type 1 = Individuals, Type 2 = Sets, Type 3 = Sets of sets, Type 4 = Sets of sets of sets, etc. etc.[83] What enables the theory to preempt Vicious Circles is that the same sort of hierarchy can be applied to propositions—e.g., Type x = some entity of some particular Type; Type $(x + 1)$ = some proposition about that entity; Type $(x + 2)$ = some proposition about that proposition about the entity; and so on. (N.B. that for Russell 'proposition' can mean either a natural-language sentence or a formal/mathematical assertion like '$a \in A$'.[84]) And

[82] **IYI** Yes: we're now coming back around to the abstract existence and denotation questions posed in §1.

[83] **IYI** If you think you can see the ghost of Aristotle's Third Man hovering around the Theory of Types, you are not mistaken. Many of the foundational problems in set theory end up looping back to Greek metaphysics.

[84] **IYI** N.B. also that Russell's arguments for the connection between the metaphysical Typology of entities/abstractions and the semantic Typology of entities/statements/metastatements are lengthy and complex, but they do exist—it's not like he's positing all this out of nowhere.

the big rule is that, where *m* and *n* are integers, a proposition or set of Type *n* cannot be applied to another proposition/set of Type *n*, but only to a proposition/set of some Type *m* where $m < n$.

As far as our Story's concerned, the Theory of Types can be seen as a perfect example of trying to legislate one's way out of a paradox. The theory does indeed offer a 'solution' to Russell's and Cantor's crunchers—that is, it gives an account of what the paradoxes' illicit move is—but it's also incredibly arcane and cumbersome, and ultimately as damaging to math as Poincaré's impredicativity thing. Quick example: Since rational numbers are defined as ratios of integers, and surds as sets/sequences of rationals, the three kinds of numbers are of different Types, and by the theory's rules we couldn't predicate things of all three in common without endless different proofs and Levels and caveats. FYI, Russell tried to patch up some of these difficulties via what he called Axioms of Reducibility, but these were even more complicated and contrived . . . and basically his whole Typology spins off into the aether and is now of merely historical interest.[85]

If it's necessary to say once again that we're just barely skimming a turbid surface here, consider it said. Specific counter-paradox measures like Russell's and Poincaré's are part of a much larger and deeper crisis, one that predates G. Cantor but is brought to a head by his theories of ∞. Thrust, broadly stated: The paradoxes of set theory, coupled with the foundational concerns that start with Abel and

[85] **IYI** Subsequent extensions and modifications of Russellian Type-theory, by logicians like F. P. Ramsey and A. Tarski, are so nightmarishly complicated and confusing that most mathematicians will pretend they don't even hear you if you try to bring them up.

Cauchy and climax with Frege and Peano, lead directly to the great Formalist-v.-Intuitionist controversies of the early 1900s. These are controversies that we again can only trace the outlines of. Respecting infinite sets, for example, Intuitionism is rabidly anti-Cantor and Formalism staunchly pro-Cantor, even though both Formalism and Intuitionism are anti-Plato and Cantor is a diehard Platonist. Which, migrainous or not, means we're again back to metaphysics: the modern wrangle over math's procedures is ultimately a dispute over the ontological status of math entities.

There's already been some intro to Intuitionism in §6's discussions of Constructivism; Formalism is its own separate kettle. The best way to come at this might be to recast the broadly stated thrust just above, to wit: The paradoxes of set theory are part of the larger issue of the Consistency of Math, which D. Hilbert proposed as Major Problem #2[86] at the same Paris Congress where he could be seen rhapsodizing about Cantor in §1a. Hilbert's own program for reconstructing mathematics in such a way that theorems don't yield paradoxes is *Formalism*, which seeks to make the abstractness of math both total and primary. The basic idea of Formalism is to totally separate math from the world and turn it into a game. Literally. This game involves the manipulation of certain symbols according to certain rules that let you construct

[86] **IYI** meaning the second of the 10 Major Unsolved Problems that Hilbert listed at 1900's 2nd I.C.M. as crucial for math to nail down in the upcoming century—another whole long story you can find in any good math-history survey. (**IYI**$_2$ If you've learned/heard that there were really 23 Hilbert Problems, the truth is that Hilbert listed 1–10 in his Paris speech and 1–23 in the written version that came out in 1902.)

sequences of symbols from other sequences of symbols. It's 100% formal—hence the name. What the math-game's symbols mean, or whether they even denote at all, has nothing to do with it; and to say that a math entity 'exists' is merely to say that it doesn't cause a contradiction.[87] What matter are the rules, and the whole project of Formalism is *proof-theoretic*: the goal is to construct a set of axioms and rules of inference[88] from which all of math can be derived, so that the whole thing's totally deductive and rigorous and clean—as a self-enclosed game, that is.

If you have any sort of background in logic or the philosophy of math, you'll recognize that this is a radically boiled-down description of Formalism. (For one thing, Hilbert's program also involves breaking math down into Levels of reasoning somewhat like Russell's Types, with again no inter-Level propositions allowed.) You'll also probably know that the movement runs into serious problems long before Gödel's aforementioned proofs that a formal system can't be

[87] Compare this Formalist ontology to the view of Intuitionism that "[M]athematical objects are mental entities that do not exist independently of our ability to provide a proof of their existence in a finite number of steps." You can see that the two views are not all that dissimilar, especially in their rejection of the idea that math has anything to do with extramental reality—although the Intuitionists' "finite number of steps" criterion is specifically meant to outlaw entities like irrationals and infinite sets that Poincaré and Brouwer, like Kronecker before them, had metaphysical (not just procedural) problems with. (**IYI** Dr. G.'s way to contrast the two schools was to say that Intuitionism was sneaky whereas Formalism was more just crazy.)

[88] **IYI** q.v. §1c.

both Complete and Consistent[89]—like e.g. the Formalists couldn't even get basic arithmetic to be Complete and Consistent if it included multiplication as a legal operation, which is obviously a serious problem. So we don't have to talk about the philosophical impoverishment or flat-out weirdness of a referentless math-game, because Formalism couldn't even succeed on its own terms.

The most coherent and successful responses to the VC paradoxes come from within set theory itself (which by c. 1900 is a thriving field in both math and logic, thanks to guess who), and are spearheaded by Cantor's #1 follower and systematizer, Prof. E. Zermelo.[90] A result of these responses is the split of abstract set theory into two subtypes, *naïve set theory* and *axiomatic set theory.* N.S.T. is just regular Cantorian set theory with all its warts and glories, including its susceptibility to paradoxes.[91] Axiomatic set theory is an attempt to derive a

[89] These C-words should have been E.G.III'd by now. They're model-theoretic terms from logic. A system is *Complete* if and only if every last true proposition can be adduced as a theorem; it's *Consistent* if it doesn't include or entail any contradictions. There's incidentally a third, briefly aforementioned criterion called *Decidability*, which concerns whether there's a procedure/algorithm for determining, for any well-formed proposition of the system, whether or not it's true (i.e., whether it's a theorem). The three criteria are obviously interconnected, but they're also distinct in important ways; and a deductively immaculate formal system is supposed to satisfy all three . . . which Gödel pretty much showed that no significantly powerful system could, which is why he's the Dark Prince, and why pure math's been in mid-air for the last 70 years.

[90] **IYI** Dates: 1871–1953. Major paper: "Investigations of the Foundations of Set Theory" (1908). Main collaborator: A. Fraenkel, who is also Cantor's first biographer.

[91] which paradoxes many working mathematicians now don't worry too much about in the course of their day-to-day work, any more than we worry about melting through the floor when we get out of bed.

more rigorous, foundationally secure version of set theory that's got all the conceptual power of N.S.T. but is set up in such a way as to avoid gross paradox. The A.S.T. program is somewhat Formalist in spirit, and Euclidean: it's to make set theory its own independent formal system[92] with its own set of axioms that yield maximal Consistency and Completeness. As mentioned someplace already, the best-known axiomatic system is usually called *ZFS* (for Zermelo, A. Fraenkel, and T. Skolem); there's also the more restrictive von Neumann–Bernays (*VNB*) system, as well as some others, w/ various metatheoretical bells and whistles, designed by eminences like A. Tarski, W. V. O. Quine, F. P. Ramsey, & c.

As it happens, axiomatic set theory and the logic of same have had fruitful applications in everything from math's theory of real functions, analysis, and topology, to generative grammar and syntax studies in linguistics, to decision theory, algorithms, logic circuitry, halting-probabilities/'Ω-studies,' A.I., and combinatorial processing in computer science. Despite increasingly dire space limitations, it is therefore worth it to include at least a doubletime tour of the basic axiomatization that all the major systems are variations of, w/ terse and directly relevant glosses where necessary—and

[92] Probably rather than 'independent' it would be better to say 'conceptually prior to,' or 'underlying,' mathematics per se. The idea behind A.S.T. is that since set theory is the most abstract and primitive branch of math, it serves as the foundation for math's most basic concepts, such as 'number,' 'function,' 'order,' etc. Though the whole issue gets very involved—particularly the questions of set theory's relation to symbolic logic and of which one is math's real fundament—it's nevertheless true that G. Frege and G. Peano, the two most important figures in the foundations of arithmetic, both define numbers and basic math operations in terms of set theory.

of course at this late point the whole thing being skip- or skimmable at your **IYI** discretion—as follows:

Primitive Concept: The membership relation $\in$, where '$s \in S$' means object s is a member of set S.

Ax. 1: Two sets are equal if they contain the same members. (Notice it's not 'if *and only if*. . .'; this is because infinite sets and their proper subsets can also be equal.)

Ax. 2: If a and b are either objects or sets, then $\{a, b\}$ is a set.

Ax. 3: There are two variants of this one. 1st variant—For a set S and a 'definite predicate'[93] P, there exists the set S_P that contains only those $x \in S$[94] that have the property designated by P. 2nd variant—There exists a set S with the following features: (a) $\emptyset \in S$, and (b) For any x, if $x \in S$, then $\{x\} \in S$. (These are two technically distinct versions of the Limited Abstraction Principle mentioned supra. Both versions do two important things. First, they establish that the empty set exists. Second, they define and validate the set-theoretic method of *transfinite induction* and, via this method, establish the existence of a denumerably infinite set S whose members are $\emptyset$, $\{\emptyset\}$, $\{\{\emptyset\}\}$, $\{\{\{\emptyset\}\}\}$, ...[95] Whereupon if, in this set, $\emptyset$ is taken to be 1 and, for any x, $\{x\}$ equals $(x + 1)$, then S becomes the ordered set of all positive integers (which

[93] **IYI** = either a single-valued function or some natural-language predicate that's meaningful for all members of S (where 'meaningful' basically means the predicate is something you can verify as definitely T v. F for any set-member, like 'is blue' or 'weighs more than 28.7 grams' as opposed to 'is lovely' or 'tastes like chicken').

[94] Using the membership symbol in a noun phrase like this is the sexy way to say 'member of S'.

[95] So an obvious corollary to the L.A.P. is: Infinite sets exist.

happens to be very close to the way Peano's Postulates[96] generate the integers).)

Ax. 4: The union of a set of sets is itself a set. (This serves as a technical definition of 'union,' from which 'intersection,' 'Cartesian Product,'[97] etc. can be derived by logical manipulation

[96] **IYI** Here's another place where it's unclear exactly which readers will know or remember what's being tossed off. If *Peano's Postulates* are not familiar and you'd like them to be, invest 45 seconds in the following: P.'s P.s are the five basic axioms of number theory; they're how you derive the whole infinite sequence of positive integers from just two primitive concepts, which latter are (a) 'is an integer,' and (b) 'is a successor of'. In natural language, the Postulates are: (1) 1 is an integer; (2) If x is an integer, the successor of x is an integer; (3) 1 is not the successor of an integer; (4) If the successors of two numbers x and y are equal, then $x = y$; (5) If a set I contains 1, and if, for any integer x in I, the successor of x is in I, then every integer is in I. Just why Postulate (5) is the axiomatic authority behind proof by mathematical induction becomes clearer in an alternative formulation, which goes more or less: (5) If P is a certain property, and if 1 has P, and if whenever an integer x has P, the successor of x has P, then all integers have P.

(**IYI$_2$** Gorisian factoid: Though Peano does deserve 100% credit for introducing all kinds of important stuff to number- and set theory (not least the standard symbols '$\in$,' '$\cap$,' and '$\cup$'), his eponymous Postulates are a clear case of capricious math-fame, since axioms equivalent to (1)–(5) appeared in Dedekind's "Nature and Meaning of Numbers" at least two years before Peano's own *Arithmetices Principia Nova Methode Exposita* came out.)

[97] **IYI** We won't bother much with *Cartesian Products* except to say (1) that they're a specific kind of interset union involving 'ordered pairs,' which are a whole saga to themselves; and (2) that C.P.s instantiate the important principle of *Preservation of Homogeneity*, meaning that if two sets A and B both have certain special characteristics, their Cartesian Product ($A \times B$) will also have those characteristics (like if A and B are point sets and both characterize topological spaces, their C.P. will also be a topological space).

(rather the way you can define the logical connective 'and' wholly in terms of 'not-' and 'or').)

Ax. 5: The famous Power Set Axiom: For any set *S*, there exists the Power Set P(*S*) of *S*. (This one establishes the infinite hierarchy of infinite sets. Recall from §7b ff. that all set theory is trivial in the case of finite sets, w/ 'trivial' meaning you can check the veracity of any set-theoretic proposition just by looking at the members of the relevant sets. The whole point of these axioms is to be able to prove theorems that are trans-experiential, 100% abstract—just like ∞ itself.)

Ax. 6: The famous and infamous Axiom of Choice. In the nomenclature of set theory, the A.C. is: 'If *S* is a set of pairwise disjoint nonempty sets, the Cartesian Product of the members of *S*[98] is not empty; every member of this Cartesian Product is designated a selection set of *S*'. In regular English, it's that from any *S* you can construct a subset *S′* with a particular property even if you can't specify a procedure for choosing the individual members of *S′*. (Zermelo came up with the Axiom of Choice in the early 1900s. It's way too technical to try to unpack here,[99] but one important consequence of the A.C. is the *well-ordering principle*, viz. that any subset *S′* of any set *S* can be chosen and arranged in such a way that

[98] **IYI** Here 'Cartesian Product' specifically means (deep inhalation) 'the set of just those subsets of the union of all members of *S* such that each (=each subset) contains exactly one member of each set in *S*'. This sort of thing is just to let you sample the heady bouquet of real A.S.T.

[99] **IYI** Any decent mathematical-logic or set-theory text will give you a whole chapter on the Axiom of Choice and its relation to such other high-eros concepts as Russell's Multiplicative Axiom, Zorn's Lemma, the Trichotomy Principle, the Hausdorff Maximal Principle, and (no kidding) the Teichmüller-Tukey Maximal Element Lemma.

S' has a first member. We've already seen the importance of this principle in 1-1C demonstrations, e.g. the very first one about {all integers} and {all positive integers} having the same cardinality. The w.o.p. is also crucial for Cantor's proofs that $c > \aleph_0$ and $P(I) > I$, since these proofs' various arrays all obviously had to have a first element. But the Axiom of Choice was also horribly controversial (for one thing, you can understand why Intuitionists and Constructivists hated the idea that you could designate a subset without any kind of procedure for picking its members), and it remained one of the great vexed questions of set theory until (1) K. Gödel in 1940 proved the A.C.'s logical Consistency with set theory's other axioms, and then (2) Prof. P. Cohen[100] in 1963 proved the A.C.'s logical Independence from (i.e., its negation's Consistency with) set theory's other axioms, which proofs together pretty much settled the Axiom's hash.[101])

Ax. 7: This one's usually known as the Axiom of Regularity; it too has several versions. The simplest one is that whether x is an object or a set, $x \notin x$. A racier formulation is: 'Every nonempty set S contains a member x such that S and x

[100] **IYI** an American (!) whom we're also about to see in action w/r/t the Continuum Hypothesis, just below.

[101] **IYI** The proof-career of the A.C. is—surprise—a very long story; the upshot of which is captured in the following from E. Mendelson's 1979 *Introduction to Mathematical Logic* (w/ the second sentence being about as heated as a logician ever gets):

> The status of the Axiom of Choice has become less controversial in recent years. To most mathematicians it seems quite plausible and it has so many important applications in practically all branches of mathematics that not to accept it would seem to be a willful hobbling of the practicing mathematician.

have no common member.'[102] (The Axiom of Regularity sort of encapsulates Poincaré's and Russell's objections to self-reference; or at any rate it's this axiom that heads off Russell's Antinomy. It also bars formulations like 'the set of all sets' and 'the set of all ordinal numbers' and so avoids Cantor's Paradox and the soon-to-be-explained Burali-Forti Paradox. Notice it also disallows Cantor's ϕ-based proof of $P(A) > A$ in §7e. This is why there's the whole separate Power Set Axiom above, from which $P(A) > A$ can be derived without any sort of proof that violates the Axiom of Regularity. But please be informed that even with the A.R., axiomatics like ZFS can still be prone to certain *model-theoretic* paradoxes,[103] so that as of say 2000 CE there's now a whole hierarchy of axiomatizations for set theory, each with its own special immunity to paradoxes, known in the trade as *Consistency strength.* If you're interested—and because if nothing else

[102] **IYI** If you'd like to see the A.R. in 100% naked symbolism, it's $(\forall S)[(S \neq \emptyset) \rightarrow (\exists x)((x \in S) \ \& \ (x \cap S = \emptyset))]$, in which the only unfamiliar symbols might be the predicate-calculus quantifiers '$\forall S$' (which means 'For all S . . .') and '$\exists x$' (which means 'There exists at least one x such that . . .' (w/ 'exists' meaning mathematically/set-theoretically (which of course assumes that this kind of existence is distinct from some other kind(s)))).

[103] These have to do mainly with how many different valid interpretations an axiom system can have (*model* being the uptown term for a specific interpretation of what the abstract symbols and formulae really stand for). It turns out that most reasonably Complete axiomatics have a ∞ of valid models—sometimes even a nondenumerable ∞ of them—which entails enormous headaches, since systems like ZFS or Peano's Postulates are set up with fairly specific models in mind, and it's not difficult to see that, with an actual ∞ of possible models, some are going to contradict the desired ones.

their names are fun—today's main systems, listed in order of increasing Consistency strength, are: Peano's Postulates, Analytic, ZFS, Mahlo, VNB, Quinian, Weakly Compact, Hyper-Mahlo, Ineffable, Ramsey, Supercompact, and *n*-Huge.)

END S-I. §7f.

§7g. You've doubtless noticed that it's been a while since G. Cantor *même* has been mentioned and have maybe wondered where he is in all §7f's foundational roil. Poincaré's and Russell's prophylactics, Zermelo's axiomatizations, etc. are all around the early 1900s, by which time Cantor's best work is behind him and he's mostly abandoned math for the obsessive preoccupations that consumed his later years.[104] It's also now that he's in and out of hospitals all the time. The poignant irony is that it's just when Cantor's work is gaining wide acceptance and set theory is inflorescing throughout math and logic that his illness gets really bad, and there are all sorts of special conferences and awards presentations he can't go to because he's too sick.

More directly apropos is that even when Cantor first happened on his paradoxes in (probably) the 1880s, he didn't worry too much about them, or rather couldn't, because he had more pressing problems. As in mathematically. Chief among them is what's now known as the *Continuum Hypothesis.*[105] The

[104] **IYI** His two main ones were Jesus's real (biological) paternity and the Bacon-v.-Shakespeare question. By way of armchair psych, both these issues concern not just factual accuracy but the denial of credit to someone deserving. Given the amount of professional shit Cantor took, his choice of obsessions seems thus both understandable and sad.

[105] **IYI** In some texts this is referred to as the *Continuum Problem.*

C.H. gets characterized in all kinds of different ways—'Is the power of the Continuum equivalent to that of the second number class?'; 'Do the real numbers constitute the Power Set of the rational numbers?'; 'Is c the same as $2^{\aleph_0}$?'; 'Does $c = \aleph_1$?'—but here's the nub. Cantor has already proved that there's an infinite hierarchy of infinite sets and their Power Sets, and he's proved that $P(A) = 2^A$ and $2^A > A$ are theorems for infinite sets. But he hasn't yet proved just how these different results are connected. The central question is whether the $2^A > A$ thing constitutes an exhaustive law for how the transfinite hierarchy is arranged—that is, whether for any infinite set A the next larger set is always 2^A, with no intermediate ∞s between them—and thus whether this process of 'binary exponentiation' is the way you get from one infinite set to the next, just the way addition lets you get from one integer to the next. A yes to this long question is the Continuum Hypothesis. What's now regarded as the general form of the C.H. is $2^{\aleph_n} = \aleph_{n+1}$,[106] but Cantor's original version is more specific. We know that he's proved the existence and cardinalities of two distinct infinite sets, namely the set of all integers/rationals/algebraics ($=\aleph_0$) and that of all reals/transcendentals/continuous intervals and spaces ($=c$); and he's proved that $c > \aleph_0$. His own Continuum Hypothesis is that $c = 2^{\aleph_0}$, i.e. that c is actually $\aleph_1$, the very next infinite set after $\aleph_0$, with nothing in between.[107]

[106] Mathematicians who call it the Continuum Problem frame the general form as 'Does there exist a set of higher cardinality than $\aleph_n$ but lower cardinality than $P(\aleph_n)$?'

[107] **IYI** It's Cantor's specific focus on c that gave the C.H. its name.

Cantor's attempts to prove the C.H. went on through the 1880s and '90s, and there are some heartbreaking letters to Dedekind in which he'd excitedly announce a proof and then a couple days later discover an error and have to retract it. He never did prove or disprove it, and some pop-type historians think the C.H. is what really sent Cantor over the edge for good.

Mathematically speaking, the truth about the Continuum Hypothesis is more complicated than pop writers let on, because Cantor really comes upon the C.H.'s various problems through his work on ordinal numbers, which numbers' relations are rather more like the '$R = \mathfrak{D}(P', P'', P''', \ldots)$' thing of §7b, and which despite our best intentions we now have to sketch very briefly.[108] First, to save time, please recall or review §5e(1) FN 78's primer on ordinal v. cardinal integers. We're now concerned with ordinal numbers in set theory, which are a little different, and involve the concept of sets' *order-types.* Simple explanation: We know that if sets *A* and *B* have the same cardinal number, they are 1-1C-able. If this 1-1C can be carried out in such a way that the order of the members of *A* and *B* remains unchanged, then *A* and *B* are the same *order-type.* (A straightforward example of two sets with the same cardinality but different order-types was {all positive integers} and {all integers} in §7c. Remember that we

[108] **IYI** N.B. that what follows is, even by our standards, a woefully simplistic overview of Cantor's theory of ordinals—a theory that's even more complex and ramificatory than the cardinal number stuff—and the only reason we're even dipping a phalange here is that it would be more woeful still to pretend that the C.H. has only to do with the hierarchy of transfinite cardinals.

had to tinker with the latter set's order so that it would have a first member to match up with the set of positives' 1.)

You can see why this is going to be more complicated than the cardinals: we're now concerned not just with a set's number of members but with the way in which they're arranged. Or rather ways, because the possible permutations of these arrangements form a good part of the ordinal theory's meat. Which meat we will now look at, though you should be aware that there are a great many technical terms and distinctions—'ordered,' 'well-ordered,' 'partially ordered,' 'everywhere-' v. 'nowhere dense,' 'relation number,' 'enumeration theorem,' and so on—that we are going to mostly blow off.[109] Some basic facts: For finite sets, cardinality = order-type; that is, two finite sets with the same cardinal number will automatically have the same order-type. This is because there's exactly one order-type for all sets with one member, one order-type for all sets with two members, and so on.[110] The total number of possible order-types for finite sets is, in fact, the same as the cardinal number of the set of positive integers, namely $\aleph_0$. It's with infinite sets that order-types get complicated. Which should be unsurprising. Take the prenominate denumerably infinite set of all positive integers: {1, 2, 3, 4, . . .} has more than one order-type. This doesn't mean just switching certain

[109] **IYI** G. Cantor's horripilatively technical theory of ordinals and sets' order-types gets worked out mostly in two papers, "Principles of a Theory of Order-Types" (1885) and the booklet-length "Contributions to the Founding of the Theory of Transfinite Numbers" (1895).

[110] The reason that might be confusing at first is the same reason our initial explanation's "in such a way that the order of the members of *A* and *B* remains unchanged" was simplistic—order-type is *not* the same as mere arrangement. '{*a*, *b*}' and '{*b*, *a*}' are the same order-type, for instance.

chunks of numbers in the infinite sequence around, since the set will still be 1-1C-able with the original set of positive integers, even if the correspondence is something like

$$\begin{array}{cccc} 2 & 18 & 6{,}457 & 1 \ldots \\ \updownarrow & \updownarrow & \updownarrow & \updownarrow \\ 1 & 2 & 3 & 4 \ldots \end{array}$$

But if you take one of the set of integers' members and put it last—as in {1, 3, 4, 5, 6, 7, . . ., 2} you now have a totally different order-type. The set {1, 3, 4, 5, 6, 7, . . ., 2} is no longer 1-1C-able with a regularly ordered $\aleph_0$ set that has no last member and so gives you no way to arrive at anything to match up with the 2. Plus observe that in the new order-type, 2 becomes a different ordinal number: it is no longer the 2nd member of the set but rather now the last member, and it has no specific number immediately before it. Hence the comprehensive def. of *ordinal number*: It's a number that identifies where a certain member of a set appears in a certain order-type.[111]

In Cantorian set theory there are two main rules for generating ordinal numbers. (1) Given any ordinal number *n*, you can always derive the next ordinal, which is $n + 1$. (2) Given any set *N* of ordinal *n*'s ordered in an increasing sequence (e.g., the set of positive integers), you can always derive a last ordinal that's bigger than all the other *n*'s. This final ordinal technically functions as the limit of *N*'s sequence and can be

[111] **IYI** Another tailored analogy of Dr. G.'s was that cardinal numbers are like the characters in a school play and ordinal numbers are the marks they're supposed to hit in a scene, as in a play's script v. its stage directions.

written 'Lim(N)'.[112] These rules don't look too bad, but things start to get tricky when we consider not just sets of ordinal numbers but ordinal numbers *as sets*—which we can do because a basic tenet of set theory is that all math entities can be represented as sets (e.g., the transfinite cardinal '$\aleph_0$' *is* the set of cardinal numbers $\{1, 2, 3, 4, \ldots\}$; plus recall §2a's *ante rem* thing about '5' literally being the set of all quintuples). So but then just what set *is* some ordinal number n? The answer is Cantor's third big rule: for any ordinal n, $n =$ the set of all ordinals less than n; i.e., n is identified with just that set of ordinals of which it is the limit.[113] Or, in formal terms,[114] $n = \{(\forall x)x < n\}$. You can generate the whole sequence of regular integers (as either cardinals or ordinals) this way: $0 = \{(\forall x)x < 0\} = \emptyset$; $1 = \{(\forall x)x < 1\} = \{0\}$; $2 = \{(\forall x)x < 2\} = \{0, 1\}$; and so on. The ordinal number of the whole denumerably infinite set $\{0, 1, 2, 3, 4, \ldots\}$ gets symbolized by the little omega 'ω'. This transfinite ordinal is the limit of the set's members' sequence—that is, it's the very smallest number bigger than all finite integers. Another, more common way to describe ω is that it's the

[112] **IYI** We can see here some clear affinities with Cantor's theory of irrationals as limits of number-sequences (in §6e). This earlier theory was, in certain ways, the origin of his work on ordinals.

[113] **IYI** The heretofore undefined *Burali-Forti Paradox* drops right out of this definition. Consider the set of all ordinal numbers everywhere. Now consider this set's own ordinal number, which by definition will be greater than any ordinal in the set—except that set was defined as containing *all* ordinals. So either way there's a contradiction. This is a mean one, and it's the real prophylactic motive behind the Axiom of Regularity.

[114] Except q.v. §7f FN 102 for what the upcoming '$\forall x$' means—which in retrospect means that FN 102 should not have been classed **IYI**.

ordinal number of that set of which $\aleph_0$ is the cardinal number.[115]

IYI INTERPOLATION

However hard the last ¶ seemed, most everything beyond that in the theory of ordinals is so brutally abstruse and technical that we can only make some general observations. One is that the arithmetic of transfinite ordinals is different from but no less weird than that of transfinite cardinals—for example, $(1 + \omega) = \omega$, but $(\omega + 1) > \omega$ because by definition $(\omega + 1)$ is the very next ordinal after ω. Another is that, just as with the cardinal $\aleph$s, an infinite hierarchy of transfinite ordinals of infinite sets of ordinals is generatable (you might want to read that last clause over), though in this case it's a

[115] **DEFINITELY IYI** Not sure it's smart to mention this, but at least sometimes G. Cantor used '$\aleph_0$' to designate the 1st transfinite ordinal and 'ω' to designate the 1st transfinite cardinal. The strict truth is that it was really the set of all finite ordinals (which is what he really called the "first number class") that Cantor used to derive the first transfinite cardinal—which he basically did because in his theory cardinal numbers were also definable as *limit ordinals*, which concept we're not discussing because it requires a level of set-theoretic detail on the relations between cardinal and ordinal numbers that this booklet's not equipped for. We are using what's now come to be the standard symbolism, viz. '$\aleph$'s for transfinite cardinals and 'ω's for transf. ordinals; the reasoning is that this symbolism stands the best chance of being familiar to at least some readers. (N.B. that the undiluted poop on Cantorian math in all its intricacies is available in several good technical books, including Dauben's aforementioned *Georg Cantor*, Abian's aforementioned *Theory of Sets*, and E. V. Huntington's *The Continuum and Other Types of Serial Order, with an Introduction to Cantor's Transfinite Numbers*—q.v. Bibl.)

very different process from the $2^{\aleph_n} = \aleph_{n+1}$ thing. The transfinite-ordinal hierarchy is associated both with abstract entities called *epsilon numbers* and with an arithmetical operation called *tetration.* We're not getting near the former except to say that they're essentially a class of numbers such that $\omega^{\varepsilon} = \varepsilon$[116]; but tetration is simpler, and you might already be familiar with it from, say, field theory or combinatorics if you had a lot of college math. It's basically exponentiation on acid. The 4th tetration of 3 is written '${}^{4}3$' and means $3^{(3^{(3^3)})}$, which $= 3^{(3^9)}$, which $= 3^{19,683}$, which you are hereby dared to try to calculate. The technical connection between tetration, transfinite ordinals, and epsilon numbers is the fact that $\varepsilon_0 = {}^{\omega}\omega$, which isn't all that important. But if you can conceive, abstractly, of a progression like ω, $((\omega+1)$, $(\omega + 2), \ldots, (\omega + \omega))$, ω^2, ω^{ω}, ${}^{\omega}\omega$, ${}^{\omega^{\omega}}\omega, \ldots$, then you can get an idea—or at any rate an 'idea'—of the hierarchy and the unthinkable heights of ordinal numbers of infinite sets of infinite sets of the ordinals of infinite sets it involves. End general obs.

END IYI I.

All right, so the specific way that Cantor runs up against the Continuum Hypothesis concerns ordinals and order-types. We've seen that there's more than one order-type for infinite sets, as with {1, 2, 3, 4, . . .} v. {1, 3, 4, . . ., 2} a few ¶s

[116] **IYI** and that they are related to the Weierstrassian epsilons of §5e only in the sense that they're created by a similar 'there exists . . . such that'–type definition—e.g., the first ordinal number k such that $\omega^k = k$ is designated 'epsilon 0' or 'ε_0'.

back. There are, in fact, a ∞ of different order-types for any infinite set; and what Cantor proves[117] is that the set of possible order-types for a denumerably infinite set is itself *non*-denumerable. This means that there's yet another distinct way to generate an infinite hierarchy of infinite sets—if *S* is some denumerably infinite set, then *Z* is the nondenumerably infinite set of possible order-types of *S*, and Z' will be the set of possible order-types of *Z*, and . . ., and away we go. (Actually, to call the different processes for deriving ∞-hierarchies 'distinct' is a little misleading, because in truth they're related in all kinds of ways. The math of these relations is beyond our technical scope here, but you can get at least some notion of the connections from the technical definition Cantor gives of set *Z* (keeping in mind that 'number class' really refers to sets of ordinals), viz.: "The second number class $Z(\aleph_0)$ is the entirety $\{\alpha\}$ of all the order-types α of well-ordered sets of the cardinality $\aleph_0$.")

It doesn't have to get that deep, though. Leaving transfinite ordinals like ω out of it, we can still see a marked and surely not coincidental similarity among (1) ***c*** as the set of all real numbers (v. $\aleph_0$'s rationals); (2) $\aleph_1$ as the Power Set of $\aleph_0$, i.e. as $2^{\aleph_0}$; (3) *Z* as the set of all order-types of $\aleph_0$. The real problem is that Cantor can't prove a certain crucial connection between these three identities. You'll recall from a couple pages back that Cantor's original C.H. is that (1) and (2) are the same, that $c = 2^{\aleph_0} = \aleph_1$, and that there's no kind of intermediate-size ∞ between $\aleph_0$ and ***c***. We are now set up to understand at least roughly how relation (3) is involved here. In the later §s of "Contributions . . . Numbers"—through a

[117] **IYI** in §15 of the prenominate "Contributions . . . Numbers."

process of profoundly, unsummarizably technical reasoning—Cantor is able to deduce two big things: (a) that there is no way that c is $> 2^{\aleph_0}$, and (b) that if there *does* exist any infinite set that's greater than $\aleph_0$ but smaller than c, this set has got to be the nondenumerable set Z, a.k.a. the second number class. It is big thing (b) that informs his main attack on the C.H., which consists in an attempt to show that relations (2) and (3) above are actually the same—that is, if Cantor can prove that $Z = 2^{\aleph_0}$, then by (b) it will be provable that there exists no intermediate set between $\aleph_0$ and c, which will entail that $c = \aleph_1$. It's specifically this $Z = 2^{\aleph_0}$ that he couldn't prove. Ever. Despite years of unimaginable noodling. Whether it's what unhinged him or not is an unanswerable question, but it is true that his inability to prove the C.H. caused Cantor pain for the rest of his life; he considered it his great failure. This too, in hindsight, is sad, because professional mathematicians now know exactly why G. Cantor could neither prove nor disprove the C.H. The reasons are deep and important and go corrosively to the root of axiomatic set theory's formal Consistency, in rather the same way that K. Gödel's Incompleteness proofs deracinate all math as a formal system. Once again, the issues here can be only sketched or synopsized (although this time Gödel is directly involved, so the whole thing is probably fleshed out in the Great Discoveries Series' Gödel booklet).

The Continuum Hypothesis and the aforementioned Axiom of Choice are the two great besetting problems of early set theory. Particularly respecting the former, it's important to distinguish between two different questions. One, which is metaphysical, is whether the C.H. is true or false. The other is whether the C.H. can be formally proved or disproved from the axioms of standard set theory.[118] It's the second question that has been definitively answered, over a period of several decades, by K. Gödel and P. Cohen, to wit:

1938—Gödel formally proves that the general form of the Continuum Hypothesis is Consistent with the axioms of ZFS—i.e., that if the C.H. is treated as its own axiom and added to those of set theory no logical contradiction can possibly result.

1963—In one of those out-of-nowhere *coups d'éclats* that pop scholars and moviemakers love, a young Stanford prof. named Paul J. Cohen proves that the *negation* of the general C.H. can also be added to ZFS without contradiction.[119]

[118] These two questions collapse into one only if either (1) formal set theory is an accurate map/mirror of the actual reality of ∞ and ∞-grade sets, or (2) formal set theory *is* that actual reality, meaning that a given infinite set's 'existence' is all and only a matter of its logical compatibility with the theory's axioms. Please notice that these are just the questions about the metaphysical status of abstract entities that have afflicted math since the Greeks.

[119] **IYI** If set and proof theory weren't so incredibly esoteric, there would already have been a big-budget movie about Cohen's proof and the stories surrounding it, which math historians love and you can find in myriad sources. What's apposite for us are some eerie parallels with G. Cantor. For one thing, Cohen's background is in functional and harmonic analysis, areas that involve both differential equations and Fourier Series—meaning that he too comes to set theory from pure analysis. It gets eerier. Cohen's Ph.D. dissertation (U. of Chicago, 1958) is entitled *Topics in the Theory of Uniqueness of Trigonometric Series.* Plus, just as Cantor had invented entirely new, Diagonal and 'ϕ'-type set-theoretic proofs, so too Cohen invents a whole new proof-technique known as *forcing,* which is prohibitively high-tech but in some ways resembles a sort of Manichean math induction where you're requiring the '$n = 1$' and 'k' cases to take one of only two possible values. . . . Which may not make sense but isn't all

These two results together establish what's now known as the *Independence of the Continuum Hypothesis*, meaning that the C.H. occupies a place rather like the Parallel Axiom's[120] w/r/t the rest of Euclidean geometry: it can be neither proved nor disproved from set theory's standard axioms.[121] Plus you'll recall from the previous § that Gödel and Cohen are able to derive pretty much the same results for the Axiom of Choice so vital to Cantor's various Diagonal proofs—Gödel proving that the Axiom isn't disprovable in ZFS/VNB, Cohen that it isn't provable in ZFS/VNB.[122] There are, as was mentioned, alternative axiom systems in which the C.H. and A.C. are provable/disprovable (e.g., Quinian set theory is set up in

---

that vital here—what's Hollywoodesque is that Cohen gets turned on to set theory, invents and refines his proof-method, and proves the C.H.'s Independence all within a single year.

[120] **IYI** Q.v. §§ 1d and 5b.

[121] This kind of Independence (which can also be called Undecidability) is a big deal indeed. For one thing, it demonstrates that Gödel's Incompleteness results (as well as A. Church's 1936 proof that 1st-order predicate logic is also Undecidable) are not just describing theoretical possibilities, that there really are true and significant theorems in math that can't be proved/disproved. Which in turn means that even a maximally abstract, general, wholly formal mathematics is not going to be able to represent (or, depending on your metaphysical convictions, contain) all real-world mathematical truths. It's this shattering of the belief that 100% abstraction = 100% truth that pure math has still not recovered from—nor is it yet even clear what 'recovery' here would mean.

[122] **IYI** Plus, in yet another '63 proof, Cohen was able to show that even if the Axiom of Choice is added to the other axioms of ZFS, the general-form Continuum Hypothesis still isn't provable—which establishes that the A.C. and C.H. are also Independent of each other, which again knocked the socks off the math world.

such a way that the Axiom of Choice is prima facie contradictory), although many of these Consistency-enhanced systems use 'set' in ways that are awfully different from Cantor et al.'s original definitions.

The Continuum Hypothesis remains alive in other ways. It is, for instance, the motive cause behind several different theoretical axiomatizations and extensions of set-theoretic models in which the C.H. and various equivalents are assumed[123] to be proved or disproved. These speculative systems are among the most hyperabstract constructs in modern math, involving rarefied terms like 'Cantorian' v. '1st-Order Universes,' 'constructible' v. 'nonconstructible sets,' 'measurable cardinals,' 'inaccessible ordinals,' 'transfinite recursion,' 'supercompletion,' and many others that are fun to say even if one has no clear idea what they're supposed to denote.[124] By way of closure, the more important thing for us to consider is how the C.H.'s unprovability bears on the other big question—whether the Hypothesis is in fact *true.* There are, not surprisingly, *n* different possible views on this. One kind of Formalist take is that various axiomatizations have various strengths and weaknesses, that the C.H. will be provable/disprovable in some and Undecidable in others, and that which system you adopt will depend on what your particular purpose is. Another,

[123] Here 'assumed' = in a speculative, what-if way. (**IYI** Factoid regarding the same clause's 'various equivalents': W. Sierpinski's 1934 *Hypothèse du Continu* lists over 80 mathematical propositions that either equal or reduce to the general-form C.H.)

[124] **IYI** A good deal of contemporary set theory seems to involve arguing about what these theoretical terms mean and just when and why they do (=mean what they mean, if anything (and, if not anything, then what that nothing might mean (and so on))).

more strictly Hilbertian response will be that 'true' in this context can't really mean anything except 'provable in ZFS,' and thus that the C.H.'s logical Independence from ZFS means it's literally neither true nor false.[125] A pure Intuitionist is apt to see the whole mess of paradox and unprovability in set theory as the natural consequence of allowing fuzzy and unconstructive concepts like sets, subsets, ordinals, and of course actual-type ∞ into math.[126]

But it is the mathematical Platonists (sometimes a.k.a. Realists, Cantorians, and/or Transfinitists) who are most upset by the C.H.'s Undecidability—which is interesting, since the two most famous modern Platonists are G. Cantor and K. Gödel, who together are at least two-thirds responsible for the whole nonplus. The Platonic position here is nicely summarized by Gödel, writing about his own and Cohen's proofs of the C.H.'s Independence:

> Only someone who (like the Intuitionist) denies that the concepts and axioms of classical set theory have any meaning could be satisfied with such a solution, not someone who believes them to describe some well-determined reality. For in reality Cantor's conjecture must be either true or false, and its undecidability from the axioms as known

[125] **IYI** Again, you can see how this Formalist view also incorporates elements of Intuitionism, the most obvious of which is the willingness to bag LEM.

[126] **IYI** L. E. J. Brouwer's pronouncement on the whole Consistency-v.-Undecidability thing in set theory is the very Aristotelian-sounding "Nothing of mathematical value will be attained in this manner; a false theory which is not stopped by a contradiction is nonetheless false, just as a criminal policy unchecked by a reprimanding court is nonetheless criminal."

today can only mean that these axioms do not contain a complete description of reality.

That is, for a mathematical Platonist, what the C.H. proofs really show is that set theory needs to find a better set of core axioms than classical ZFS, or at least it will need to add some further postulates that are—like the Axiom of Choice—both "self-evident" and Consistent with classical axioms. If you're interested, Gödel's own personal view was that the Continuum Hypothesis is false, that there are actually a whole ∞ of Zeno-type ∞s nested between $\aleph_0$ and c, and that sooner or later a principle would be found that proved this. As of now no such principle's ever been found. Gödel and Cantor both died in confinement,[127,128] bequeathing a world with no finite circumference. One that spins, now, in a new kind of all-formal Void. Mathematics continues to get out of bed.

[127] **IYI** Hilbert didn't go out easy either. Brouwer and Russell, on the other hand, both ended up living so long they practically had to be dispatched with clubs.

[128] **IYI** As of this writing, P. J. Cohen is the Marjorie Mhoon Fair Professor of Quantitative Science at Stanford.

Grateful acknowledgment is made to the following for their help: Classena Bell, Jesse Cohen, Mimi Bailey Davis, Jon Franzen, Bob Goris and the H. of D., Rochelle Hartmann, Rich Morris, Erica Neely, Joe Sears, Stephen Stern, John Tarter, Jim Wallace, Sally Wallace, and Bob Wengert.

It goes without saying that the author is solely responsible for any errors or imprecisions in this booklet.

Scholarly Boilerplate

CITATIONS FOR QUOTED
AND/OR CRIBBED MATERIAL
(FULL SPECS FOR ALL RELEVANT SOURCES
ARE IN THE BIBLIOGRAPHY)

§1a. p. 5 "One conclusion appears . . . " = Bell, pp. 521–522.

§1a. p. 6 "the finest product . . . purely intelligible" = Hilbert, "*Über das Unendliche,*" p. 197.

§1a. p. 6 "Poets do not . . . " = Chesterton as cribbed from Barrow, p. 171.

§1a. p. 6 "In the late nineteenth . . . " = Amir D. Aczel, *The Mystery of the Aleph: Mathematics, the Kabbalah, and the Search for Infinity,* Four Walls Eight Windows, 2000.

§1b. p. 8 "But what . . . " = Boyer, p. 596.

§1b. p. 9 FN 5 "In the beginning . . . " = Russell, *Mysticism and Logic,* p. 60.

§1b. p. 10 "One of the great . . . " = Kline, p. 29.

§1b. p. 10 "What has escaped . . . " = Saussure, p. 23.

§1b. p. 18 "No data processing . . . " = Klir and Yuan, pp. 2–3.

§1c. p. 20 "The philosophical doctrines . . . " = Kline, p. 175.

§1c. p. 20 "[T]he infinite is nowhere . . . " = Hilbert, p. 191.

§1c. p. 25 " 'The certainty of . . . " = Hardy, *Apology,* p. 106.

§1d. p. 36 "an element of . . . " = semicribbed, w/ light editing, from Nelson, p. 330.

§1d. p. 40 "The fundamental flaw . . . " = Cantor, "*Über die verschiedenen Standpunkte in bezug auf das aktuelle Unendliche,*" in *Gesammelte*

Abhandlungen. Edited translation cribbed from Dauben, *Georg Cantor,* p. 125.

§1d. p. 41 FN 27 "I entertain no doubts . . . " = Cantor in Dauben's "Georg Cantor and Pope Leo XIII," p. 106.

§1d. p. 41 FN 27 "The fear of infinity . . . " = Cantor, *Gessamelte Abhandlungen,* p. 374. Edited translation cribbed from Rucker, *Infinity and the Mind,* p. 46.

§1d. p. 41 "We may say . . . " = Hardy, *Apology,* p. 89.

§1d. p. 42 "the attributes 'equal' . . . " = G. Galilei, p. 32.

§2a. p. 44 "the unlimited substratum . . . " = Edwards, v. 3, p. 190.

§2a. p. 44 "[T]he essence of the infinite . . . " = Aristotle, *Physics,* Book III, Ch. 7, 208a.

§2a. p. 45 FN 3 "Beauty is the first test . . . " = Hardy, *Apology,* p. 85.

§2a. p. 48 "In this capricious . . . " = Russell, *Mysticism and Logic,* p. 76.

§2a. p. 52 "Zeno was concerned . . . " = Ibid., p. 77.

§2a. p. 62 FN 20 "I believe that the numbers . . . " = Hermite in Barrow, p. 259.

§2b. p. 63 'What exactly . . . ?' and 'Of what sort . . . ?' = semicribbed from Edwards, v. 3, p. 183.

§2b. p. 63 "For there are two senses . . . " = Aristotle, *Physics,* Book VI, Ch. 2, 233a.

§2b. p. 63 "So while a thing . . . " = Ibid.

§2b. p. 65 "[I]t is as that . . . " = Aristotle, *Metaphysics,* Book IX, Ch. 6, 1048a–b.

§2b. p. 67 "In the direction . . . " = Aristotle, *Physics,* Book III, Ch. 6, 207b.

§2b. p. 67 "One thing can . . . " = Ibid, 206b.

§2c. p. 81 FN 39 "Since, in proving . . . " = M. Stifel, *Arithmetica Integra,* 1544, as cribbed from Kline, p. 251.

§2d. p. 82 "Magnitudes are said . . . " = Euclid in Heath, v. 2, p. 114.

§2d. p. 85 "If from any magnitude . . . " = Euclid in Boyer, p. 91.

§3a. p. 93 "The existence of . . . " = *Summa Theologiae,* I.a., 7.4. Translation semicribbed, w/ editing, from Rucker, *Infinity and the Mind,* p. 52.

§3a. p. 96 'Florentine architect . . . viewer-eye-level' = little more than a light massage of Honour and Fleming, pp. 319–320.

§3a. p. 98 "It was largely . . . " = Boyer, p. 322.

§3a. p. 99 "reduced to actuality . . . " = Ibid., p. 329, w/ some editing.

§3a. p. 100 "when we attempt . . . " = G. Galilei, p. 31.

§3a. pp. 100–101 "we must say that . . . finite quantities" = Ibid., p. 32.

§3a. p. 101 "When, after a thousand-year . . . " = Tobias Danzig, *Number: the Language of Science,* as cribbed from Seife, p. 106.

§3a. p. 104 "It is the contrast . . . " = Berlinski, p. 130.

§3a. p. 105 "to which the progression . . . " = Gregory of St. Vincent, *Opus Geometricum,* as cribbed from Kline, p. 437.

§3b. p. 107 "the story this world . . . " = Berlinski, p. xi.

§3b. p. 108 "As science began . . . " = Kline, p. 395.

§3b. p. 108 "Infinitely large quantities . . . " = Ibid., p. 393.

§3c. p. 118 "The term 'analysis' . . . " = Clapham, p. 6.

§4a. p. 130 FN 11 "the sum of . . . " = Boyer, p. 403.

§4a. pp. 131–132 Exhibit 4a and parts of '$A = x^2 \ldots z = 2x$' = citable resemblance to Lavine, pp. 16–17.

§4a. p. 133 FN 15 "It is useful . . . " = Leibniz as cribbed from Kline, p. 384.

§4b. p. 138 FN 27 "Nothing is easier . . . " and "[H]e who can digest . . ." = Berkeley in *Works,* v. 4, pp. 68–69.

§4b. p. 138 FN 27 "the true metaphysics . . . " and "A quantity is . . ." = d'Alembert, *Encyclopédie.* Translation semicribbed from Boyer, p. 450, w/ editing.

§4b. p. 144 FN 35 'which solution . . . ' = little more than a slight rewording of Sainsbury, p. 21.

§5a. p. 147 "a virtual fusion . . . " = Kline, p. 395.

§5a. p. 148 FN 5 "[It used to be that] geometry . . . " = Clairaut as cribbed from Kline, pp. 618–19, w/ editing.

§5a. pp. 151–152 Differential-equation stuff in ¶ that starts with 'That may not have been . . .' = at least semi-semicribbed from Berlinski, pp. 230–231.

§5a. p. 152 FN 13 "A variable quantity becomes infinitely small . . . " and "A variable quantity becomes infinitely large . . . " = Cauchy, *Cours* *d'analyse algébrique.* Translation cribbed, w/ some editing, from Kline, p. 951.

§5b. p. 161 'freely drawn' = phrase cribbed from Lavine, p. 30.

§5b. p. 163 "raised more questions . . . " = Dauben, *Georg Cantor*, p. 7.

§5b. p. 165 "[M]athematicians began to . . . " = Kline, p. 947.

§5b. p. 165 "It would be a serious . . . " = Cauchy, Intro to *Cours.* Translation cribbed from Kline, ibid., w/ light editing.

§5b. p. 165 "There are so very . . . " = Abel, pp. 263–264.

§5c. p. 169 "Too little information . . . " = Dauben, *GC*, p. 288.

§5c. p. 170 "Georg Woldemar had a thoroughly . . . " = Bell, as paraphrased by Dauben (who absolutely loathes Bell), *GC*, p. 278.

§5c. p. 170 "a sensitive and gifted . . . " = Dauben, ibid.

§5c. p. 170 FN 42 "Had Cantor been brought up . . . " = Bell, p. 560.

§5c. p. 170 "Whereas Georg Ferd. . . . " = Dauben, *GC*, p. 278.

§5c. p. 171 "took a special . . . " = Ibid., p. 277.

§5c. p. 171 "poor health" = Grattan-Guinness, "Towards a Biography," p. 352.

§5c. p. 172 FN 43 " . . . I close . . . to success!" = Dauben, *GC*, pp. 275–277.

§5d. p. 175 "*y* is a function . . . " = Dirichlet, "*Über die Darstellung. . . .*" Translation cribbed from Kline, p. 950.

§5e. p. 180 "From [Zeno] to our own . . . " = Russell, *Mysticism and Logic*, p. 77.

§5e. p. 182 "[T]he history of mathematical . . . " = Grattan-Guinness, *From the Calculus to Set Theory*, p. 132.

§5e. p. 183 '$f(x)$ is *continuous* at some point x_n . . . ' = semicribbed from Kline, p. 952.

§5e. p. 183 "$f(x)$ will be a *continuous* . . . " = Cauchy, *Cours*, p. 35.

§5e. p. 188 '*limit point* . . . of the set' = semicribbed from Lavine, p. 38.

§5e. p. 188 FN 72 "1 is a limit point . . . " = Ibid.

§5e. p. 192 FN 78 "In this theory . . . " = Russell, *Principles of Mathematics*, p. 304.

§5e. p. 196 "by saying that *the sum* . . . " = Courant and Robbins, p. 64.

§5e. p. 196 "This 'equation' does not . . . " = Ibid.

§6a. p. 197 "It was Weierstrass . . . " = Kline, p. 979.

§6a. p. 197 "The irrationals which . . . " = Bell, p. 419.

§6a. p. 198 "And that theory promptly . . . " = Lavine, p. 38.

§6a. p. 198 "highly illogical; for if . . . " = Russell, *Principles of Mathematics*, p. 278.

§6a. p. 200 "In stimulating conversation . . . " = translation cribbed, w/ some editing, from Bell, p. 519.

§6a. p. 202 "Of the greatest importance . . . " = Dedekind, p. 8.

§6a. p. 202 FN 5 "[I]t will be necessary . . . " = Ibid., p. 5 (syntax *sic*).

§6b. p. 203 "I protest against . . . " = Gauss as cribbed from Kline, p. 993.

§6b. p. 203 "In speaking of arithmetic . . . " = Dedekind, p. 31.

§6b. p. 204 "Numbers are free . . . " = Ibid.

§6b. p. 204 "My own realm . . . " = Ibid., p. 64, w/ light editing.

§6b. p. 205 "The idea of considering . . . " = Cantor, "Foundations of the Theory of Manifolds," pp. 75–76.

§6c. p. 207 "By vague remarks . . . " = Dedekind, pp. 10–11.

§6c. p. 207 "[T]he majority of my readers . . . " = Ibid., p. 11.

§6c. p. 208 "If all points . . . " = Ibid.

§6c. p. 212 "In this property . . . " = Ibid., p. 15, w/ editing.

§6d. p. 214 "[I]f one regards . . . " = Ibid., w/ editing.

§6d. p. 215 "All the more beautiful . . . " = Ibid., p. 38 (punctuation *sic*).

§6d. p. 216 FN 22 "What use is your . . . ?" = semicribbed from Kline, p. 1198.

§6d. p. 218 "hearty thanks . . . of continuity" = Dedekind, p. 3, w/ editing.

§6e. p. 221 'an infinite sequence . . . arbitrary *m*' = semicribbed from Kline, pp. 984–985.

§7a. p. 228 "The modern theory . . . " = Lavine, p. 41

§7a. p. 228 "But the uninitiated . . . " = Russell, *Mysticism and Logic*, p. 82.

§7a. p. 231 FN 9 "To every number . . . " = Cantor, *Gessamelte Abhandlungen*, p. 97. Translation cribbed from Dauben, *GC*, p. 40.

§7a. p. 234 FN 15 "Cantor stressed . . . " = Ibid.

§7a. p. 235 FN 16 "A collection into a whole . . . " = translation cribbed from Edwards, v. 7, p. 420.

§7a. p. 236 'finite-' v. 'infinite iterations' = phrase blatantly cribbed from Lavine, pp. 40–41.

§7b. p. 242 "We see here a dialectic . . . " = Cantor, *Gessamelte Abhandlungen*, p. 148. Translation cribbed from Grattan-Guinness, *From Calc to S.T.*, p. 191.

§7b. p. 243 '$1 + 2 + 3 + \ldots (\aleph_1)^{\aleph_0} = \aleph_1$' = semicribbed from Edwards, v. 7, p. 422.

§7b. p. 244 "The transfinite numbers . . . " = Cantor, *Gessamelte Ab.*, pp. 395–396. Translation cribbed from Dauben, *GC*, p. 128, w/ editing. All emphases *sic.*

§7b. p. 244 "their existence is . . . " = Cantor as summarized by Dauben, *GC*, p. 127.

§7b. p. 245 FN 34 "In particular, in introducing . . . " = Cantor, *Gessamelte Ab.*, p. 182.

§7b. p. 245 FN 34 "For mathematicians . . . " = Dauben, *GC*, p. 128.

§7b. p. 245 FN 35 "The concept of . . . " = Fraenkel in Edwards, v. 2, pp. 20–21.

§7c. p. 249 '$C = 1 \ldots \frac{n-1}{2}$' = semicribbed from Bunch, p. 115, w/ substantial editing.

§7c. p. 250 'relations of magnitude' = phrase cribbed from Courant and Robbins, p. 79.

§7c. p. 256 "There are the sets . . . " = Benacerraf and Putnam, p. 410.

§7c. p. 257 "A reality completely . . . " = Poincaré as cribbed from Barrow, p. 276.

§7c. p. 257 "Despite their remoteness . . . " = Gödel, "What is Cantor's Continuum Problem?" p. 270.

§7d. p. 260 'since every geometer . . . in that space' = semicribbed from Grattan-Guinness, *From Calc to S.T.*, p. 188.

§7d. p. 262 FN 56 "pulling the terms . . . variables" = Cantor as summarized by Dauben, *GC*, p. 55.

§7d. p. 263 "what wonderful power . . . " = Cantor in Cavaillès, p. 234. Edited translation semicribbed from Dauben, *GC*, pp. 55–56, w/ the variable '*n*' substituted for Cantor's original 'ρ'.

§7d. p. 264 "This [G.C.'s] view seems opposed . . . " = Ibid., and ditto w/r/t the '*n*' and 'ρ'.

§7d. p. 265 "It seems utterly . . . " = Du Bois-Reymond, from French translation of *Die allgemeine Funktiontheorie* (1882), w/ edits semicribbed from Kline, p. 998.

§7e. pp. 269–272 Arrays and parts of 'Call the set of all integers *I* . . . a true 1-1C between *I* and P(*I*) is impossible' = citable resemblance to Hunter, pp. 22–23.

§7f. p. 283 FN 87 "[M]athematical objects are . . . " = Nelson, p. 229.

§7f. p. 288 'If *S* is a set of pairwise . . . ' = semicribbed from Edwards, v. 7, p. 425, w/ editing.

§7f. p. 289 FN 101 "The status of . . . " = Mendelson, p. 213.

§7f. p. 289 'Every nonempty set *S* . . . ' = semicribbed from Edwards, v. 7, p. 426, w/ editing.

§7f. p. 290 FN 102 '$(\forall S)[(S \neq \emptyset)$ . . . ' = cribbed from Mendelson, p. 213, w/ minor changes in symbolism.

§7g. p. 292 FN 106 'Does there exist . . . ?' = semicribbed from Edwards, v. 2, p. 208.

§7g. p. 298 'The 4th tetration . . . it involves' = suspicious resemblance to Rucker, *Infinity and the Mind*, pp. 74–75.

§7g. p. 299 "The second number class . . . " = Cantor, *Contributions . . . Numbers*, p. 44.

§7g. p. 302 FN 121 'theoretical possibilities' = phrase cribbed from Bunch, p. 158.

§7g. p. 304 FN 126 "Nothing of mathematical . . . " = Brouwer as cribbed from Barrow, p. 217.

§7g. p. 304 "Only someone who . . . " = Gödel, "What . . . Problem?" p. 268. Edited version semicribbed from Barrow, p. 264.

Bibliography

BOOKS

Niels H. Abel, *Oeuvres complètes*, v. 2, Johnson Reprint Corp., 1964.

Alexander Abian, *The Theory of Sets and Transfinite Arithmetic*, W.B. Saunders Co., 1965.

Howard Anton, *Calculus with Analytic Geometry*, John Wiley & Sons, 1980.

John D. Barrow, *Pi in the Sky: Counting, Thinking, and Being*, Clarendon/Oxford University Press, 1992.

Paul Benacerraf and Hilary Putnam, Eds., *Philosophy of Mathematics: Selected Readings*, Prentice-Hall, 1964.

J. A. Benardate, *Infinity: An Essay in Metaphysics*, Clarendon/Oxford University Press, 1964.

Eric T. Bell, *Men of Mathematics*, Simon & Schuster, 1937.

David Berlinski, *A Tour of the Calculus*, Pantheon Books, 1995.

Max Black, *Problems of Analysis*, Cornell University Press, 1954.

Carl Boyer, *A History of Mathematics*, 2nd ed. w/ Uta Merzbach, John Wiley & Sons, 1991.

T. J. I. Bromwich and T. MacRobert, *An Introduction to the Theory of Infinite Series*, 3rd ed., Chelsea Books, 1991.

Bryan H. Bunch, *Mathematical Fallacies and Paradoxes*, Van Nostrand Reinhold Co., 1982.

Georg Cantor, *Contributions to the Founding of the Theory of Transfinite Numbers*, trans. P. E. B. Jourdain, Open Court Publishers, 1915; Reprint = Dover Books, 1960.

Georg Cantor, *Transfinite Numbers: Three Papers on Transfinite Numbers from the Mathematische Annalen*, G. A. Bingley Publishers, 1941.

Georg Cantor, *Gesammelte Abhandlungen mathematischen und philosophischen Inhalts* (= Collected Papers). Eds. E. Zermelo and A. Fraenkel. 2nd ed., G. Olms Verlagsbuchhandlung, Hildesheim FRG, 1966.

Augustin-Louis Cauchy, *Cours d'analyse algébrique*, = v. 3 of Cauchy, *Oeuvres complètes*, 2nd ed., Gauthier-Villars, Paris FR, 1899.

Jean Cavaillès, *Philosophie mathématique*, Hermann, Paris FR, 1962 (has French versions of all the important Cantor-Dedekind correspondence on pp. 179–251).

Nathalie Charraud, *Infini et Inconscient: Essai sur Georg Cantor*, Anthropos, Paris FR, 1994.

Christopher Clapham, Ed., *The Concise Oxford Dictionary of Mathematics*, 2nd ed., Oxford University Press, 1996.

Paul J. Cohen, *Set Theory and the Continuum Hypothesis*, W. A. Benjamin, Inc., 1966.

Frederick Copleston, *A History of Philosophy*, v. I pt. II, Image Books, 1962.

Richard Courant and Herbert Robbins (Revised by Ian Stewart), *What Is Mathematics? An Elementary Approach to Ideas and Methods*, Oxford University Press, 1996.

Joseph W. Dauben, *Georg Cantor: His Mathematics and Philosophy of the Infinite*, Princeton University Press, 1979.

Richard Dedekind, *Essays on the Theory of Numbers*, trans. W. W. Beman, Open Court Publishing Co., 1901; Reprint = Dover Books, 1963.

Paul Edwards, Ed., *The Encyclopedia of Philosophy*, 1st ed., v. 1–8, Collier MacMillan Publishers, 1967.

P. E. Erlich, Ed., *Real Numbers, Generalizations of the Reals, and Theories of Continua*, Kluwer Academic Publishers, 1994.

J.-B. Joseph Fourier, *Analytic Theory of Heat*, Dover Books, 1955.

Abraham Fraenkel, *Set Theory and Logic*, Addison-Wesley Publishing Co., 1966.

Galileo Galilei, *Dialogues Concerning Two New Sciences*, Dover Books, 1952.

Alan Gleason, *Who Is Fourier?* Transnational College of LEX/Language Research Foundation, 1995.

Kurt Gödel, *The Consistency of the Axiom of Choice and of the Generalized Continuum-Hypothesis with the Axioms of Set Theory*, Princeton University Press, 1940.

Ivor Grattan-Guinness, Ed., *From the Calculus to Set Theory*, Gerald Duckworth & Co., London UK, 1980.

Leland R. Halberg and Howard Zink, *Mathematics for Technicians, with an Introduction to Calculus*, Wadsworth Publishing Co., 1972.

Michael Hallett, *Cantorian Set Theory and Limitation of Size*, Oxford University Press, 1984.

G. H. Hardy, *Divergent Series*, Oxford University/Clarendon Press, 1949.

G. H. Hardy, *A Mathematician's Apology*, Cambridge University Press, 1967/1992.

T. L. Heath, *The Thirteen Books of Euclid's Elements*, v. 1–3, Dover Books, 1954.

Hugh Honour and John Fleming, *The Visual Arts: A History*, Prentice-Hall, 1982.

Geoffrey Hunter, *Metalogic: An Introduction to the Metatheory of Standard First Order Logic*, University of California Press, 1971.

E. V. Huntington, *The Continuum and Other Types of Serial Order, with an Introduction to Cantor's Transfinite Numbers*, Harvard University Press, 1929.

Stephen C. Kleene, *Introduction to Metamathematics*, Van Nostrand, 1952.

Morris Kline, *Mathematical Thought from Ancient to Modern Times*, v. 1–3, Oxford University Press, 1972.

George J. Klir and Bo Yuan, *Fuzzy Sets and Fuzzy Logic: Theory and Applications*, Prentice-Hall, 1995.

Shaughan Lavine, *Understanding the Infinite*, Harvard University Press, 1994.

Paolo Mancuso, Ed., *From Brouwer to Hilbert: The Debate on the Foundations of Mathematics in the 1920s*, Oxford University Press, 1998.

W. G. McCallum, D. Hughes-Hallett, and A. M. Gleason, *Multivariable Calculus* (Draft Version), John Wiley & Sons, 1994.

Richard McKeon, Ed., *Basic Works of Aristotle*, Random House, 1941.

Elliott Mendelson, *Introduction to Mathematical Logic*, 2nd ed., D. Van Nostrand Co., 1979.

Robert Miller, *Bob Miller's Calc I Helper*, McGraw-Hill, 1991.

David Nelson, Ed., *The Penguin Dictionary of Mathematics*, 2nd ed., Penguin Books, 1989.

Theoni Pappas, *Mathematical Scandals*, Wide World Publishing, 1997.

Henri Poincaré, *Mathematics and Science: Last Essays*, trans. J. W. Boldue, Dover Books, 1963.

W. V. O. Quine, *Set Theory and Its Logic*, Belknap/Harvard University Press, 1963.

Georg F. B. Riemann, *Collected Mathematical Works*, Dover Books, 1953.

Rudy Rucker, *Infinity and the Mind*, Birkhäuser Boston, Inc., 1982.

Bertrand Russell, *Introduction to Mathematical Philosophy*, Allen and Unwin, London UK, 1919.

Bertrand Russell, *Mysticism and Logic*, Doubleday Anchor Books, 1957.

Bertrand Russell, *Principles of Mathematics*, 2nd ed., W. W. Norton & Co., 1938.

Gilbert Ryle, *Dilemmas: The Tarner Lectures 1953*, Cambridge University Press, 1960.

R. M. Sainsbury, *Paradoxes*, Cambridge University Press, 1987.

Ferdinand de la Saussure, *Cours de linguistique générale* (R. Engler, Ed.), Harrasowitz, Wiesbaden FRG, 1974.

Charles Seife, *Zero: The Biography of a Dangerous Idea*, Viking Press, 2000.

Waclaw Sierpinski, *Hypothèse du Continu*, Monografie Matematyczne, Warsaw PL, 1934.

Patrick Suppes, *Axiomatic Set Theory*, D. Van Nostrand Co., 1965.

University of St. Andrews, *MacTutor History of Mathematics Web Site*: 'www.groups.dcs.st-and.ac.uk/~history'.

I. M. Vinogradov, Ed., *Soviet Mathematical Encyclopedia*, v. 9, Kluwer Academic Publishers, 1993.

Eric W. Weisstein, *CRC Concise Encyclopedia of Mathematics*, CRC Press, 1999.

Hermann Weyl, *Philosophy of Mathematics and Natural Science*, Princeton University Press, 1949.

ARTICLES & ESSAYS

George Berkeley, "*The Analyst*, Or a Discourse Addressed to an Infidel Mathematician Wherein It is Examined Whether the Object, Principles, and Inferences of the Modern Analysis are More Distinctly Conceived, or More Evidently Deduced, than Religious Mysteries and Points of Faith. 'First Cast the Beam Out of Thine Own Eye; and Then

Shalt Thou See Clearly to Cast Out the Mote Out [*sic*] of Thy Brother's Eye,'" in A. A. Luce, Ed., *The Works of George Berkeley, Bishop of Cloyne*, Thomas Nelson & Sons, 1951.

Jorge L. Borges, "Avatars of the Tortoise," in D. Yates and J. Irby, Eds., *Labyrinths*, New Directions, 1962, pp. 202–208.

Luitzen E. J. Brouwer, "Intuitionism and Formalism," trans. A. Dresden, *Bulletin of the American Mathematical Society* v. 30, 1913, pp. 81–96.

Georg Cantor, "Foundations of the Theory of Manifolds," trans. U. R. Parpart, *The Campaigner* No. 9, 1976, pp. 69–97.

Georg Cantor, "*Principien einer Theorie der Ordnungstypen*" (="Principles of a Theory of Order-Types"), mss. 1885, in I. Grattan-Guinness, "An Unpublished Paper by Georg Cantor," *Acta Mathematica* v. 124, 1970, pp. 65–106.

Joseph W. Dauben, "Denumerability and Dimension: The Origins of Georg Cantor's Theory of Sets," *Rete* v. 2, 1974, pp. 105–135.

Joseph W. Dauben, "Georg Cantor and Pope Leo XIII: Mathematics, Theology, and the Infinite," *Journal of the History of Ideas* v. 38, 1977, pp. 85–108.

Joseph W. Dauben, "The Trigonometric Background to Georg Cantor's Theory of Sets," *Archive for the History of the Exact Sciences* v. 7, 1971, pp. 181–216.

H. N. Freudenthal, "Did Cauchy Plagiarize Bolzano?" *Archive for the History of the Exact Sciences* v. 7, 1971, pp. 375–392.

Kurt Gödel, "Russell's Mathematical Logic," in P. A. Schlipp, Ed., *The Philosophy of Bertrand Russell*, Northwestern University Press, 1944.

Kurt Gödel, "What is Cantor's Continuum Problem?" in Benacerraf and Putnam's *Philosophy of Mathematics*, pp. 258–273.

Ivor Grattan-Guinness, "Towards a Biography of Georg Cantor," *Annals of Science* v. 27 No. 4, 1971, pp. 345–392.

G. H. Hardy, "Mathematical Proof," *Mind* v. 30, 1929, pp. 1–26.

David Hilbert, "*Über das Unendliche*," *Acta Mathematica* v. 48, 1926, pp. 91–122.

Leonard Hill, "Fraenkel's Biography of Georg Cantor," *Scripta Mathematica* No. 2, 1933, pp. 41–47.

Abraham Robinson, "The Metaphysics of the Calculus," in J. Hintikka, Ed., *The Philosophy of Mathematics*, Oxford University Press, 1969, pp. 153–163.

Rudolf v. B. Rucker, "One of Georg Cantor's Speculations on Physical Infinities," *Speculations in Science and Technology*, 1978, pp. 419–421.

Rudolf v. B. Rucker, "The Actual Infinite," *Speculations in Science and Technology*, 1980, pp. 63–76.

Bertrand Russell, "Mathematical Logic as Based on the Theory of Types," *American Journal of Mathematics* v. 30, 1908–09, pp. 222–262.

Waclaw Sierpinski, "*L'Hypothèse généralisée du continu et l'axiome du choix*," *Fundamenta Mathematicae* v. 34, 1947, pp. 1–6.

H. Wang, "The Axiomatization of Arithmetic," *Journal of Symbolic Logic* v. 22, 1957, pp. 145–158.

R. L. Wilder, "The Role of the Axiomatic Method," *American Mathematical Monthly* v. 74, 1967, pp. 115–127.

Frederick Will, "Will the Future Be Like the Past?" in A. Flew, Ed., *Logic and Language*, 2nd Series, Basil Blackwell, Oxford UK, 1959, pp. 32–50.

Index

(Note: numbers in *italics* refer to illustrations and photos)